THE
CHINA
BLUEPRINT

THE
CHINA
BLUEPRINT

A BIOGRAPHY &
CHINA STRATEGY BLUEPRINT

BRAD GOOD

This book is dedicated to Manman. I love and miss you. I'm sorry things worked out the way they did. I hope you and your family are safe and happy.

Table of Contents

SECTION I:
A BIOGRAPHY

Introduction

Chinese leaders have a vivid recollection of what is known as the "Century of Humiliation." This period, spanning from the mid-19th century to the mid-20th century, was marked by foreign domination, military defeats, territorial losses, and political chaos. Key events, such as the Second Opium War, the First and Second Sino-Japanese Wars, and the Chinese Civil War, profoundly shaped both China's national identity and the perspectives of its leaders. The concept of humiliation has become deeply ingrained in Chinese consciousness, serving as a foundational narrative for the Chinese Communist Party (CCP). When President Xi Jinping speaks of achieving the "great rejuvenation of the Chinese nation," he is addressing the rectification of this historical grievance. This narrative is crucial to understanding modern Chinese identity.

In light of this historical context, *The China Blueprint* presents a distinctive approach: for the first time, America is not asking anything from China. Any strategy that America adopts must foster strategic autonomy while respecting China's collective memory of humiliation and its strong desire to prevent similar injustices from reoccurring. *The China Blueprint* is designed to be broad, robust, and strategic, aiming to resonate with both the Chinese populace and the CCP leadership. Achieving this is no small feat but is essential for success.

The primary aim of this book is to propose a much-needed strategy for the United States to effectively engage with China. Unlike other works that dwell on China's transgressions or instill fear about its intentions—an approach that has become all too common—this book seeks to outline a justified strategy accompanied by actionable initiatives.

China has often been misunderstood in the international arena. For instance, when a surveillance balloon traversed American airspace, it led to sanctions against Chinese component manufacturers. Similarly, the use of child labor in Xinjiang prompted restrictions on certain products entering the US market. As fentanyl continues to claim over 150 American lives daily, the response has been focused on restricting its entry from Mexico.

In essence, while many have focused on symptomatic issues, few have developed a strategy that addresses their root causes. In Chinese philosophy, this is encapsulated in the phrase *zhìbiǎo, búzhìběn*—treating symptoms rather than underlying problems. It is imperative that we recognize these ongoing issues and concentrate on their *fundamental* causes to devise an effective strategy.

If America continues without a well-considered approach and remains fixated on surface-level threats, such negligence will only perpetuate China's well-documented atrocities. This represents one of America's most significant challenges of this generation. My hope is to illuminate how these issues can be addressed. Ultimately, this book should serve as a wake-up call for politicians, media figures, and others that the United States must formulate a comprehensive strategy with appropriate initiatives and personnel to tackle the root causes of China's actions. The primary source of these challenges is clear: the senior leadership of the Chinese Communist Party. A detailed account of China's atrocities can be found in *Section VII: The China Proclamation*.

Before delving into this proposed strategy and its initiatives, I would like to share some personal experiences from my time in China that will provide valuable context for *The China Blueprint*. I began living in Beijing in 1988, a time when China was on the cusp of significant change. Since then, I have witnessed its transformation firsthand; I now reside in Taipei.

My journey in understanding China began long before the conception of *The China Blueprint*, and it is these experiences that have shaped my unique perspective on US-China relations.

CHAPTER 1

First Move to Hong Kong, 1986

While at UC Berkeley, I immersed myself in the works of James Clavell, becoming particularly enchanted by *Tai-Pan*. This novel chronicles the rise of the world's largest Far East trading company and its supreme leader, Dirk Struan. The story follows Struan's adventures as he trades opium between India and China, drawing inspiration from the real-life Jardine Matheson, which emerged in 1832 as the most successful opium trading company, headquartered in Hong Kong.

My fascination with Hong Kong and China motivated me to study Mandarin at UC Berkeley. I believed that working in Hong Kong would offer far more excitement than remaining in America. I anticipated China's eventual opening, which would present lucrative business opportunities, and so, with strong conviction, I enrolled in Chinese language classes in 1984.

In class, I was the only white student among my peers. The instructor was exceptional; each day, he posed questions using new vocabulary, requiring us to be prepared or risk embarrassment. I became adept at creating flashcards for new words and phrases, meticulously labeling and reviewing them before tests and each new semester. My classmates were primarily from Taiwan and Hong Kong; their parents had often insisted on using only English at home to help them assimilate into American society. Now adults, they sought to reconnect with their Chinese heritage, driven by a desire to understand their roots and cultural identity. However, their appearance and familial ties did not translate into an advantage in class.

Years later, I discovered that one of my classmates hailed from one of Hong Kong's most esteemed families. Only elite families from Hong Kong or Taiwan could afford to send their children to foreign schools; other immigrant families were mostly residents of California.

Fifteen days after graduation in December 1986, I packed two suitcases and moved to Hong Kong. The Peninsula Hotel, which opened in 1928 and overlooks Victoria Harbour, epitomizes colonial luxury and oriental chic. Instead of staying there, I chose the more economical YMCA hotel next door. Although rundown, its location facing Hong Kong Island was ideal, and I was fortunate enough to have a harbor view.

I knew no one in Hong Kong except for one contact. I needed to find a job and build an entirely new life, which felt overwhelming. Fear was not an option, though; I was determined to succeed. My father was a respected OB/GYN in Santa Monica, while my mother was an artist. They neither opposed nor fully understood my decision.

Growing up in Pacific Palisades, Los Angeles, I had kept in touch with an old girlfriend. Before leaving for Hong Kong, I visited her family; her father, Morgan Harris, was a top executive at Korn Ferry, a prestigious executive search firm. Morgan kindly provided me with the name and phone number of their top manager in Hong Kong. With cautious optimism, I set out to meet Peter for our arranged meeting.

His office was located in the St. George's Building, a charming colonial structure built in 1908 near the harbor on Hong Kong Island. As I entered his office, in a dark blue suit, Peter greeted me warmly: "Brad, it's nice to meet you. Welcome to Hong Kong!"

"Thanks, Peter. It's great to meet you," I replied.

His secretary approached and asked if she could get me something to drink; I requested a black coffee.

After taking a sip of my coffee, Peter asked how he could assist me today—a phrase that caught me off guard, as it seemed too formal for our situation—he didn't start with small talk or mention Morgan.

I responded candidly: "Well, I just graduated from UC Berkeley and Morgan Harris suggested I contact you regarding potential job opportunities here in Hong Kong."

"Brad," he said bluntly. "You do not speak Chinese and have no experience. Go home."

Peter was the head of the most prestigious executive search firm in Hong Kong; he could easily connect me with opportunities, if he chose to. All I sought was a chance to learn and prove myself. Something felt amiss—

perhaps he had political disagreements with Morgan—but my instincts told me how to handle such situations.

Maintaining my dignity, I stood up. "Thanks for your time, Peter."

"Nice to meet you, too," he replied as I walked out.

* * *

I returned to my hotel room and changed into smart casual attire before heading over to the Peninsula Hotel's concierge. An attentive lady greeted me warmly: "How may I help you?"

The contrast between her welcoming demeanor and Peter's earlier coldness struck me.

"I'm looking for a list of foreign companies in Hong Kong."

"Ah! Our business center likely has some listings," she said. "However, I recommend visiting the American Chamber of Commerce for a more comprehensive list."

"Thank you! That's exactly what I need. Could you please write down their address in Chinese for my driver?"

She wrote down the address on a Peninsula Hotel taxi card and handed it to me. As I exited the hotel and handed the card to the doorman who flagged down a taxi, my mother's words echoed in my mind: *Every cloud has a silver lining.*

*   *   *

The list from the American Chamber of Commerce included all its members, along with CEOs' names and office phone numbers. I began making calls, occasionally reaching senior executives directly with my pitch: "I'm a recent UC Berkeley graduate who has just moved to Hong Kong. I'd love to learn more about the region—can I drop by for a chat?"

I aimed to present myself as eager to learn and contribute, viewing this as an excellent chance to network and understand regional developments; little did I know how many doors I'd need to knock on for a work opportunity.

On my eighth attempt, I successfully arranged a visit with Jardine Matheson's subsidiary—Jardine Engineering. Mike Battisson met me in the lobby. He had a strong English accent and, after an hour-long chat, he introduced me to the company president. They needed someone for marketing projects and offered local wages of about $1,000 per month, along with private Chinese tutoring lessons included. I started work the following week, just two weeks after arriving in Hong Kong. I began my first position filled with excitement about being employed in this vibrant city.

I didn't realize it then, but upon reflection, this was a defining moment in my life. I had taken a risk, even against the advice of an expert. This would become a pattern: charting the course of my life without regard for those who might not share my vision. It was the first step on a path that would lead me through the complexities of Hong Kong, Taiwan, and ultimately, China.

*   *   *

My private tutor, originally from Beijing, came to my office three times a week. During the ninety-minute sessions, he would test me, correct my tones, and make me write essays. He ensured every single tone was correct. The importance of this was emphasized one evening when we were looking out the office window. I commented: "*hěn dūo dàsha.*" I meant to say: "There are a lot of buildings." Although we were overlooking buildings and the comment was in context, he did not understand. I had said "*dàsha*" instead of "*dàshà*". I learned a valuable lesson—you cannot simply speak *most* of your tones correctly and expect Chinese to understand you. It is difficult to memorize every tone, but this is not a valid excuse for being sloppy when speaking.

Every night and weekend, I studied Chinese, forfeiting a social life. Work went well, and I liked all the staff in the department who were from England and France. I was the token Yank. They made fun of me in a nice way. They were educated and committed to the company.

After several months, Mike called me into his office. "We'd like you to spend some time in Taiwan and write up a plan for us to expand business there."

I blinked. "What's your thinking?"

"It's a growing market. Jardine's has an office there, but we do not. Simon is the head. We'll help organize some meetings for you. Then write up a plan."

"Sounds good." I was twenty-two years old with no experience. I spent the next few weeks learning about what Jardine Engineering was really good at. Where did it make the most money? I also met with all the department heads and tried to figure out what the company was really good at. I thought that was a good start.

Little did I know my venture to Hong Kong would so soon introduce me to Taiwan, opening my eyes to completely new experiences and challenges in a place I had only vaguely imagined.

*   *   *

In 1987, I landed in Taipei. I was placed in a suite in a three-star hotel. That was to be my home for six months. Simon, with his strong Scottish accent, was funny and delightful. Several nights he took me out and showed me the more colorful sides of Taipei. I learned a lot about a part of Taiwan society. As the head of Jardine in Taipei, he was rather senior within the organization.

They organized meetings. My knowledge skyrocketed. I informed the people I met with what Jardine was good at and asked questions about how we might be able to work together. During one of the meetings with the president of a major company, I heard mice above us, running back and forth. It was embarrassing. But, it was 1987 in Taipei. The president showed no notice of mice having sex in the ceiling above. The poor construction quality was evident in many buildings at that time.

Each morning I would go for a jog. I'd make my way across a bridge under which water flowed. Some days I had to turn back; the raw industrial waste in the river made me gag. Taiwan had not yet taken care of a number of serious environmental issues.

I visited the National Palace Museum. When Chiang Kai-shek was fighting the communists, they moved 2,972 crates of artifacts from the Forbidden City in Beijing to Taiwan. At any given time, the National Palace Museum displays less than one percent of the 700,000 items collected.

At that time, the main night attraction was a place called "Snake Alley". It was a lively area with food stalls where you could eat snake meat, soups, and blood. Despite the possibility of salmonella and other parasites, the area was a great attraction to foreigners. Right nearby was one of Taipei's red-light districts. Together, these two places were, at that time, one of the most vibrant areas in Taipei. Now, public attitudes about snake eating and prostitution have driven away those two pastimes. It was exciting to have seen that time in Taipei's history when it was wild and a little bit crude.

Before 1987, there was strict separation from China. Taiwan operated under the Three Noes Policy: No contact, No negotiation, and No compromise. After the new president lifted martial law that year, there was greater freedom in Taiwan, and some engagement started. I remember this being a big deal since there were family members who were separated and could finally reunite.

Hong Kong's *South China Morning Post* featured people who, after four decades, could finally visit relatives. Some veterans who retreated to Taiwan were permitted to meet with relatives in Hong Kong, but not China. There were two "Noes" that persisted: No direct phone calls between Taiwan and China, and No direct flights.

I wrote up my findings and recommendations and returned to Hong Kong. It was good to be back home and see my friends in the office. Mike thought highly of my work. It was then that I looked at him and said, "I can't survive in Hong Kong on $1,000 USD a month." This was true. I had been dipping into my savings while being very careful spending money.

This initiated a series of events over the next two months. They could not pay me a reasonable salary, given the company's remuneration schedule. The president and Mike petitioned the Jardine Matheson Human Resources head to let me into Jardine's "Carter Program." Members of this program needed to be interviewed by top company management. Program members were nearly always the sons of lords or significant Jardine Matheson stockholders. They were all from the UK. Fewer than two dozen were part of the program. There were no Americans. To their glee (and mine), interviews were arranged.

Jardine Matheson, at that time, had 72,000 staff and over 160 years of history. The company had financial services, property investments and developments, motor vehicle companies, and retail. In 1987, it was still one of the most prominent companies in Hong Kong.

As I prepared for my interview, I couldn't help but feel a mix of excitement and nervousness. This was a pivotal moment in my career, and I had to make a strong impression.

*　*　*

I went up the elevator to the 52nd floor of the Jardine House building. The building had distinct round windows which made it recognizable from nearly anywhere in Hong Kong. I remembered the roots of the company: opium. That was a long time ago. Ironically, many of the storage facilities along the harbor were now ultra-prime real estate. Who would have thought?

Simon Keswick was blood-related to the founder of Jardines and was its Tai-pan, or "big boss". He was Scottish and had studied at England's finest educational institutions, including Cambridge. Brian Powers was the American co-CEO from James D. Wolfensohn, an esteemed New York boutique investment banking firm. I met them on the top floor in an expansive office full of comfort and luxury.

I glanced at the round windows. I had only ever viewed them from outside and I couldn't help but feel like this was my first "Alice in Wonderland" experience.

Simon and Brian were friendly and asked me about where I was from and why I was in Hong Kong. I was straightforward and looked them in the eye. I had the privilege of coming from a good family; my mother had forced me to go to cotillion for four years. I was lucky enough to know how to have a cordial semi-superficial conversation but look sincere.

After about ten minutes I was done. And, it was successful.

Finally, my living situation improved. Being part of the program had some advantages; I was able to select furniture and received a furnishing allowance. I had been living on the island of Discovery Bay, which required taking a thirty-minute boat ride to and from work. It was fun, though. I met a buddy named Trey who would become a lifelong friend. But, now that my salary had increased, I moved to Hong Kong Island for convenience.

I met others who were part of the Carter Program. Their first question was always, "What does your dad do?" Some invited me to lunch at their club. We'd eat lunch and then they'd go swimming. During the week! It seemed silly and irresponsible to me. I had stuff to do.

*   *   *

Jardine Engineering arranged a training seminar in nearby Shenzhen for a bunch of staff. This was in late 1987. We jumped on a bus and drove through the border, into the city. What I saw then was clear. There were farms and rice paddies everywhere. There were zero factories or skyscrapers. Shenzhen was a rural village with about one million people.

In 1987 no one talked about the business they were doing in China. That's because they weren't doing any business; China had not opened up. Largely, China was a dust bin. People still wore Mao suits. They were poor. They were like the people who now live in North Korea.

Years later, in 1992, Deng Xiaoping toured Shenzhen to reinforce his reforms and emphasize the need to open up. Now Shenzhen has over 18 million people and is ranked as a global first-tier city. Its GDP has passed that of Hong Kong, and its economy is one of the ten largest in the world. Freedom—the restriction from government interference—does amazing things. Deng later said that he had made a mistake and should have started with Shanghai. An insightful leader.

Hong Kong life was great. There was a rather significant gap between foreigners and locals, but they blended nicely. I did not have the opportunity to meet with the police task force head of combating the triads. Crime was not often seen but it took place and was ruthless.

Being able to communicate with locals was not always easy since they all spoke Cantonese. One morning I stopped by my building's front desk and asked the doorman, "Can you please give my apartment key to an air-conditioning repairman who will be visiting later in the morning?" The doorman did not understand my English. When I switched to Mandarin, the doorman still did not understand me, despite me showing him the key in my hand. Finally, I took out a piece of paper and hand-wrote my message with Chinese characters. He gave me a great smile and two thumbs up.

At that time in history all people across China did not understand Mandarin, but they shared a written language. When I looked at the piece of paper later, I was not proud of the unattractive Chinese characters I had written. But, I reminded myself, it did not matter how it looked—it had gotten the job done. And that I was proud of. Ironically, now most Hongkongers do speak Mandarin. Most older people learned from watching China TV programming.

*   *   *

About six months later, I came to realize that, though Hong Kong was great fun, I was not learning much. Conducting the assignment in Taiwan was remarkably challenging but had been a great learning experience. I did not see a great deal of opportunities at the time to grow in Jardines or in Hong Kong. China had not yet started to open up in earnest. Plus, I knew I needed to enhance my business acumen and language abilities to prepare for when China *did* open. So, I applied to the University of Chicago to get an MBA and a Master's in East Asian Studies. I knew I could earn a much better salary with an MBA.

After I got accepted, I told my language tutor I'd be leaving. We had become trusted friends. I had passed a company language exam and used all the financial reward to buy him a watch. I felt greatly indebted to him for fiercely correcting all my tones and guiding me in the use of the language.

He suggested I go live with his friend's family in Beijing before going to Chicago. I had only visited Beijing once before, for business. It was a dreary place. Without hesitation, though, I jumped at the opportunity.

Even after only spending two years there, I knew the vibrancy in Hong Kong was thrilling. Yet it was just a small free island connected to China. It reinforced my belief that the potential in China would be massive. I appreciated there was a great deal for me to learn before I would be prepared to operate within the country. Although my two-year stint in Hong Kong was brief, I knew the only way to truly get prepared would be to leave and then return.

CHAPTER 2

1988 Beijing

In September 1988, I landed in Beijing and entered a small airport. After collecting my luggage, I made my way to the airport's sole exit and saw Chén Liáng, my teacher's friend and former classmate. He was uncharacteristically tall, at about six feet three inches, in stark contrast to others nearby, who were shorter than me. A generation of Chinese would soon be superseded by a taller generation, due to their enriched diet. This change would take only years to have an impact, one that I would witness firsthand.

After exchanging pleasantries, we got into his car and started to make our way into the city. The road we traveled on was small, with one lane each way. Looking to the left and right, I could see parallel dirt roads where we periodically passed horse-drawn carts. Finally, we made our way to the well-known Beijing neighborhood of Tuánjíehú.

We arrived on the ninth floor and Chén Liáng introduced me to his wife, Chén Tàitài. She was a doctor, so I immediately got into the habit of calling her Chén Dàifu, out of respect. Then, I was introduced to Tiāné, their eleven-year-old daughter. Tiāné, which meant "swan," was her nickname. She was not happy to see me, as she would have to sleep on the sofa while I took over her room.

That night, a small table and even smaller chairs were taken out. We had a simple dinner of mostly vegetables, with a little bit of chicken. It became immediately clear to me that pretty much everyone was poor and simply did not have the money for a typical Western-sized meal. For the first time, I truly understood that I had grown up extremely spoiled.

Our conversations were simple; I had my big dictionary right next to me. It was of limited use and often hindered the conversation. It takes too long to look up a word and the conversation needed to move on. I explained my background and that I was on my way to the University of Chicago. I was grateful for them to let me stay. Unspoken was the agreement my tutor had made that I would pay them $14 per day; it seemed like such a small sum of money. I had learned to stay in my lane when it came to sensitive things like this.

Every morning I woke up and went for a jog. They accepted that I needed to exercise. I ran about three miles, making a broad circle of the entire neighborhood. I saw people on bicycles going to work, people talking, doing laundry, and eating. I saw real life in Beijing. Upon returning to the apartment, I found out that there was no hot water. They only had hot water one day a week. They had boiled water for me to bathe. I felt guilty for having inconvenienced them, but they genuinely did not seem to mind.

Chén Liáng took me to a nearby small police station and registered me. I remember vividly how he casually put a few cigarettes in front of the officer. I had heard it was illegal for foreigners to live with locals. Chén Liáng worked for an Italian company and evidently had an Italian passport. Either way, I later heard I was likely the only foreigner in 1988 who lived with a Chinese family.

Over the next few days, I came to understand the situation a little bit better. The electricity turned off each night at nine p.m. They had no refrigerator. It would only be ten years later, in 1999, that 70 percent of Chinese households would have a refrigerator. About thirty years later, a French photographer would find negatives of a slew of pictures of people next to their new refrigerator. A photo exhibit in Shanghai displayed this amazing period in Chinese history: well-dressed women standing proudly next to a new appliance.

I also quickly found out that within the small two-bedroom apartment, Chén Liáng had a monkey, a dog, fish, and now me, a foreigner. Chén Liáng truly was open-minded! Dogs were illegal in Beijing at that time.

After everyone has a refrigerator, owns a dog and a house, what type of country will China evolve into? It would inevitably depend upon the nature of the government and the Chinese themselves. I did not see anything then that led me to believe the outcome might be elegant or fruitful—I merely was viewing a glimpse of Beijing in 1988.

One late morning Chén Liáng took me to the Silk Market. This was an outdoor market then and had a bunch of clothing kiosks and tables. Banks offered a much less favorable exchange rate compared to the black market. At this time in history, China had two currencies, FEC and RMB. The Foreign Exchange Certificates were for use by foreigners and foreign companies; RMB were used by local Chinese. Everything cost the same, but you could get many more RMB for one USD. So, I was there to exchange $300 for RMB.

I located a money changer. After a short discussion, we agreed upon an exchange rate. Then, with magical speed, he counted out the notes. I had heard stories of cheating—money changers would count bills really fast, and it would look like the amount you agreed upon, but by using sleight of hand, the number of bills fell short when you counted for yourself.

Before handing over my $300, I asked, "May I please count for myself?"

"No."

I knew then he was trying to cheat me. I was angry and said a phrase in Chinese which was not polite. Immediately, his face turned red. His three friends suddenly appeared. After a few seconds, Chén Liáng appeared out of nowhere and pulled me away toward the car.

We never spoke of what happened. He knew I knew. I used a particular phrase in Chinese that is so serious it prompts Chinese people to fight. In the car on the way back, I made a mental note to thoroughly master Chinese swear words: when to use them and what to expect. Mostly, I learned never to use them.

Over the next few weeks, I continued my daily jogging. No longer did everyone look at me weirdly. They waved, and I shouted "Hi" or "Good morning." I had become, in a short period of time, a fixture in the neighborhood. Like everyone I passed, I was doing something.

I made up my mind to go jogging on the Great Wall. No one wanted to go with me. I arranged a car to pick me up at four-thirty a.m. It was about an hour and a half away, and I wanted to see the sunrise. This was still possible since Beijing was not yet completely polluted. There are a number of places to visit on the Great Wall; my favorite now is Jiankou, which is completely unrestored, with wild and steep sections. Usually, it is closed, but as they say, "*Zǒngshì yǒu bànfǎ*" — There is always a way.

The driver took me to the most common area where foreigners visit, the Mutianyu section. Being there so early meant there were few people. While my driver waited, I jogged, carrying nothing. I passed the odd white tourists occasionally but mostly had the entire Wall to myself. There were Japanese tourists with hiking poles along the way. It was not an easy jog. The Wall goes up and down with steps, proving to be more than just a good workout.

Nearly finished, I walked down the final steps. On the right, I saw a number of stands selling various things, so I walked over. I found a vendor selling t-shirts. There were several Great Wall shirts, and I decided it would be great to get one to commemorate my first visit.

"How much for the shirt?" I pointed to one.

"Ten US dollars," he replied.

That was not a favorable response. It was one of the most famous tourist destinations and you could supposedly see the Great Wall from outer space, but still. I gave him a smirk and said in Chinese, "How about three dollars?" I knew I could get the same shirt elsewhere for one dollar.

He responded, "Nine dollars."

Still sweating from running, I said, "I'm a student. That's way too much. How about two dollars?" I had gone down on price since he was really trying to take advantage of me.

His response was feigned anger, a typical Chinese negotiating technique. "No," he said as he shooed me away with the back of his hand, in typical Chinese fashion.

I turned and started to walk back to the car. Negotiating when you are going to get taken advantage of is silly. I didn't need that shirt. Suddenly, I felt something hit me from behind. I looked back—it was the shirt. The vendor had thrown it at me. I picked it up, walked over, and handed him three, not two, dollars. He grinned.

I asked him, "Do you have a bag?"

He enthusiastically and loudly said, "Yǒu!" Yes.

That's how a Jew and a Chinese vendor negotiated and became friends. I overpaid but not by a stupid amount. At that time in Chinese history, more than 63 percent of the population was below the poverty rate. I was interacting with many Chinese and only just starting to understand them and how they thought. I knew overpaying was okay, and maybe even good. I would not negotiate this way in other parts of the world. Going down in price, counter to expectations, has never failed me in China. Then again,

I have never negotiated for something I wasn't willing to walk away from.

I visited the Beijing Friendship Store, which was the only place where you could purchase foreign goods. There was nothing terribly appealing. They did have some foreign candy, which I bought for everyone. I still felt I was not in the good graces of Tiāné.

It was hard not to think about the impact the one-child policy would have on China. Later on, Tiāné would encounter a modified policy. Since she was a girl, her parents would be permitted to have another child. They would be allowed to try for a boy. All the Chinese I asked about the one-child policy agreed that it was a good thing, saying, "There are just too many people in China." It was their words but the central government's rationale.

I tried to explain that the government could instead grow the economy by having more favorable market policies, to which all I got were shrugs. Why talk about what you cannot control? Now China is facing a dire demographic situation, with a higher percentage of older people. The government highly discourages abortions, as opposed to forcing them previously. Now, the government no longer permits the adoption of Chinese children by Americans. Why do people and governments not understand the reality of unintended consequences?

Never ask a Chinese native a question and expect an answer that comes from their own critical thinking. They have been informed by their government of what their perspective should be on nearly everything.

At about this time I reached out to the heads of a few corporations. One of the people I met with was the CEO of Siemens, the German technology conglomerate. Over coffee he shared stories of trying to do business in China. Historically, the barriers were nearly insurmountable; companies were just starting to look at the market more closely. Interestingly, he said they estimated Siemens would make back the money it had invested in about twenty-seven years. Not my kind of payback period.

*   *   *

The following evening, we were eating dinner together. There was, as usual, not much conversation after another long day. That evening, as I finished my bowl of rice, Tiāné said to me, "You are wasteful."

I didn't really know what she was referring to and looked at her with confusion.

"You are wasteful," she repeated. "You don't finish all your rice. Rice may not be expensive, but people work hard to grow it."

She was right. I had not eaten every single grain of rice in my bowl. I thanked her for telling me. She clearly learned that from her parents, the government, or both. But there is a truth to what she said that I appreciate. The irony is that Chinese eat much less rice now, and I'm usually the only one who completely finishes his bowl. For fun, when Chinese do not finish their rice, I tell them, "You are wasteful." They remember their old lesson and feel a little bit humiliated at being told that by a foreigner. This one small piece of their past, part of their culture, is gone now.

I also visited the Forbidden City, my favorite tourist destination. I asked many Chinese: *How many emperors have lived in the Forbidden City?* No one could answer. It's twenty-four! To me, it is unfathomable that so many emperors lived in the same location. It must have amazing Fengshui! In fact, it was the center of power for over 500 years during the Ming and Qing Dynasties.

The pictures of Mao are still everywhere, including on money and even now on electronic currency. Chinese haven't the slightest idea about how many people Mao killed; the lowest estimate is 50 million. I passed the main entrance of Zhongnanhai, the central headquarters of the Chinese Communist Party. There were two slogans on the side of the entrance: "Long Live the Great Chinese Communist Party" and "Long Live the Invincible Mao Zedong Thought." I noticed another clear sign that said: "Serve the People." This one was written in Mao Zedong's calligraphy and serves to underscore the CCP's role in serving the people—the ultimate role of the CCP and China's president, though it seems that, over the years, many have forgotten.

I returned to the apartment and found Dr. Chén reading. She was always fun to see and talk with. She put the paper down and looked at me. "How is it that someone who sells vegetables on the street makes just as much money as a doctor?" I knew exactly what she was referring to and, I knew why. China was a communist country, and that stifled all sorts of good stuff. I nodded my head, not saying anything but empathizing with her.

Four months passed swiftly. I started to learn lots but knew there was so much more. Saying goodbye, I expressed my deep appreciation to Mr. and Dr. Chén. I looked to Tiāné, who she seemed genuinely pleased I was departing. She too had been a wonderful help to me. Speaking with an eleven-year-old was almost at my language level. Yet I knew I had much room to improve my Mandarin.

Tiāné is now forty-eight years old. I can only imagine how she might have grown up.

Visiting China in 1988, I witnessed the stark realities of life in Beijing—the dire circumstances many faced, but also the resilience and potential of its people. The open-mindedness of Chén Líang and the sharp mind of Chén Dàifu reinforced my belief that China had immense untapped potential.

CHAPTER 3

University of Chicago

Obtaining two master's degrees—an MBA and a master's in East Asian Studies—would take three long years. Every day I had language class. The teacher was an older man from Beijing, which gave me some degree of reassurance. There were so many different accents throughout China and I had already become biased toward the more formal language spoken in Beijing and on national TV. Interestingly, Beijing was the only place in the country where another dialect was not spoken.

Just like at UC Berkeley, I was the only white person in class. I marveled that no one thought the language was worth learning. To help ensure I had everything handled, I hired a tutor once a week to review all my new words and anything else I might need.

Every morning prior to classes, I would review my flashcards in the business school lounge. My business school friends became accustomed to seeing me with stacks of white cards. Sometimes in classes when I got bored, I'd subtly review my cards. I figured I was a customer paying for each class; it was my prerogative to do what I wanted as long as I did not disturb others.

In 1991, the professors at the University of Chicago were familiar with Japan and the concept of *Kanban*, or lean manufacturing. China had not come up on their radar yet. I was convinced that they could not identify where China was on a map. It didn't matter. Compared to learning Chinese, the business classes were easy.

Just months after arriving, students began thinking about what type of summer internship they wanted. Chicago, being a finance school, meant lots of people focused on banking. Consulting and marketing were also big. I began looking at the career office and secured an interview with Booz Allen, for a summer associate position in Singapore. The interview was held by phone. I knew enough about Hong Kong, Taiwan, and China to sound intelligent and, two weeks later, they gave me a call to offer me the position. I was excited since all the consulting companies combined had offered a total of only five positions in Asia that year.

The summer in Singapore unexpectedly transformed into a stay in Surabaya, Indonesia. Booz Allen had dispatched me as part of a team to evaluate a clove cigarette manufacturer's distribution capabilities. During my assessment, I toured their facilities, including one where two thousand women hand-rolled cigarettes. The sight was so striking that I hastily retreated, acutely aware of the workers' curious glances at my uncommon Western presence.

Indonesians exude extraordinary warmth, yet they possess a volatile side best avoided when provoked. One morning, en route to the office, we received word of a strike. As we passed the factory, we witnessed hundreds of Indonesian women in protest. I silently was happy for not being the negotiator in that tense situation.

A friendship with one of the company's staff led to an invitation to their private island. Far from luxurious—except for the tennis courts—the island offered a rustic charm. As night fell, our group of ten had authentic Indonesian food and drink before sleeping on a series of sofas in the open air. In this intimate setting, I overheard two individuals conversing in Mandarin. Their Chinese was far from flawless, but I understood every word. Speaking with them later sparked a deeper appreciation for the Chinese diaspora across Indonesia and broader Asia. The conversation served as a poignant reminder of China's far-reaching influence beyond its borders.

Chinese have dispersed across the globe, particularly in Asia. Their impact on business has been profound and disproportionate to their numbers. In Indonesia alone, Chinese Indonesians control approximately 70 percent of the market value of listed companies.

This economic influence, while impressive, also hints at potential social and political tensions that could shape Indonesia's future—and cause problems with the indigenous population. This, I later learned, was especially the case in Malaysia. Local companies were limited in the number of Chinese they could hire. Actually, there is strong affirmative action for the *bumiputra* enshrined in the Malaysian Constitution.

The summer with Booz flew by, and I returned to quickly settle back into my routine. More flashcards. Again, the only white student. I started learning Chinese cursive writing. This was a complete waste of time but a mandatory part of the advanced Chinese course. I had to start all over, making flashcards for cursive writing. To this day, I have still not read a single note in cursive Chinese. Mobile phones and using Pinyin to text Chinese had not yet been invented, but would be vastly more useful.

*   *   *

As another summer was soon to arrive, I applied and was admitted to Middlebury College in Vermont. Their language program was known as the best in the United States, and in fact, the CIA used to send people there.

Middlebury College was one of the most beautiful places in the world. Zhou Zhiping, a Princeton University-based professor, was the head of the amazing Chinese program. We had to learn about one hundred words a night. Each day of class, we would turn in homework and take a Chinese test, both of which were quickly corrected and returned to us the same day.

The rule was that only Chinese could be spoken during the summer. If you spoke English, you would be thrown out, though none of us tested the school's resolve in this area. I loved it. Professor Zhou had created the best program I could have imagined. His accompanying teachers were superb—friendly, motivational, and strict.

Back in Chicago, I began my last year. One year seemed like a lot, but I began my constant routine of powering through my flashcards, morning and night. I also had to do a Master's Thesis. To me, this was just a long paper. I met with my advisor and agreed upon my topic: Guangdong Province would follow the same developmental path as Taiwan. I knew Taiwan had zoomed ahead developmentally. It was 1992 and seemed like an easy topic—review a profile of Taiwan and compare it to Guangdong, and compare the statistics of both over time.

The problem I encountered was that all the statistics were in Chinese, and their reliability was questionable at best. As it turned out, even Chinese Premier Li Keqiang harbored doubts about their accuracy. Years later, the "Li Keqiang Index" emerged (dubbed by *The Economist*), combining railway freight traffic, electricity consumption, and total bank loans as alternative economic indicators. This unorthodox approach to measuring economic activity speaks volumes about the challenges in obtaining reliable data in China.

The use of such an index surprised me. How could a nation of China's economic stature lack a more dependable measure of its GDP? My research into Chinese statistics revealed firsthand their deficiencies, stemming from the country's transition from a Soviet-style system to a more Western-oriented one.

Local governments, driven by political pressures and incentives, often overpromised on their economic performance. The central government, in turn, massaged the data to paint a rosier picture of the nation's economic health. This systemic issue of data manipulation created a fog of uncertainty around China's true economic condition.

Could implications of this statistical sleight of hand extend far beyond academic interest, potentially influencing global economic policies and investment decisions? I began to feel this was in China's DNA.

I turned in the thesis two weeks before it was due. "Good job," my advisor said. "Great topic and analysis. It gets a B+ but, if you fixed a few things, I'll bring it to an A."

I said, "That's okay. I'll take the B+."

That year I graduated two weeks early so I could take Professor Zhou's similar Chinese program at Beijing Normal University in Beijing. It didn't work out too well, though. I had to share a room and the weather during the summer in Beijing was very hot, with no air-conditioning. To make matters worse, there was not a single seat available in the library.

Professor Zhou wanted my class to start reading Chinese literature. My goal in learning Chinese was not to study Chinese history or culture; it was to communicate. I told him that instead of doing his lesson, I thought I might write a letter to the *People's Daily*, the dominant newspaper in China. At this time, I had a computer and had brought a printer. I got to work putting together an opinion letter, using pinyin to type characters on the computer. This was a new capability that only recently became available.

I printed out a version and went over it with one of the teachers. They were there to help and readily offered suggestions. This was the first article I had written and it was for a Chinese publication. Two weeks later, I sent the letter to the newspaper. Three weeks later, they printed it. I had written something I knew the paper would be receptive to printing—I wrote about Chinese wanting to move to the United States. I discussed why they wanted to move and what they would encounter if they did.

I showed Professor Zhou the article in the newspaper. He saw my byline and was amazed. He did not expect that I could appreciate what a hugely famous Chinese newspaper might be willing to print, and then to craft such an article. Chinese institutions are predictable. He did not know that, but I did. There is always some motive behind an organization and the individuals who run it; the key is understanding that motive and what drives key individuals, what they want and what they fear.

The people and the program were outstanding, but the environment was not optimal. Living in Beijing again, five years from my previous stint, was only a little bit edifying. Lots of new smaller buildings had sprouted up with street-level retail shops. The growth was indeed substantial. It was apparent that people, for the first time, were allowed to make money and keep a good portion of it.

One early evening, a friend and I made our way to a nearby dumpling restaurant. We sat outside, enjoying a rare breeze. While we ate, a bicycle pulling a cart alongside passed by. On the cart were two young men from the countryside. You could tell by their messed-up hair, their Mao-style thick shirts they wore, and their tanned faces.

I called to them in Chinese. "Hi."

Their eyes widened as they looked up at my friend and me.

"We are not aliens," I said. "By the way, what is the name of this dish we're eating?" I pointed down to the dish, hoping they would come over to take a look.

They literally jumped off the cart and ran away. Welcome to Beijing in 1992. They had never before seen a *wàiguórén*, a foreigner. It scared them and they had run away. Considering that nearly all foreigners lived in Shanghai and Beijing, this meant that the vast majority of the Chinese population had not been seen by those outside of these major cities.

I left weeks early, without my completion certificate. I could not get myself to study Chinese literature. I did not need any more certificates. I booked a flight to Hong Kong, took what I had, and left. It was time to get a real job.

My intensive language study and the start of on-the-ground experience in China had laid a good foundational basis. The experiences in Hong Kong, Chicago and Beijing—from poring over statistics for my thesis to crafting articles for the newspaper to living with a family—were merely the beginning.

To truly operate effectively in China's complex business and political landscape, I needed to develop a nuanced understanding of the country's economic systems, power structures, and unwritten rules. The challenges I faced deciphering Chinese statistics had illuminated the gap between official narratives and on-the-ground realities.

This realization shaped my approach to future engagements with China. It instilled in me a healthy skepticism toward anything official, and a commitment to seeking multiple perspectives. Moreover, it underscored the importance of building a network of trusted local contacts who could provide insights beyond what was publicly available. In reflection, *this* is when I started to get to know China. Unlike what others said, I was not a "China Expert"—I was a novice.

CHAPTER 4

1993 Hong Kong

Hong Kong was even busier than before. Everyone seemed taller, reflecting the enriched diet of the most recent generation. Southern Chinese, often viewed as shorter, were dispelling this notion. The belief that Chinese wouldn't like pizza was proven wrong as Pizza Hut thrived in Hong Kong and later throughout China.

Finding a job after obtaining two master's degrees proved less stressful. I secured a position at A.T. Kearney, which included a $10,000 signing bonus. My role involved working with a $500 million investment fund focused on the automotive component parts industry in China, the first and largest of its kind in modern Chinese history.

Traveling throughout China for meetings, I further honed my Chinese language skills and learned to interpret subtle cues in communication. People in different regions spoke Mandarin with various accents, and I had to pay close attention to their tone of voice and body language to gauge their sincerity. This experience provided valuable insights into China's diverse regions and cultures.

During winter visits, I encountered cold taxis, as drivers avoided using heaters to save gas. Most were old Volkswagen Santanas, manufactured since 1983. Over 3.5 million were produced, but the ones I rode in seemed ancient and of poor quality.

Traveling around China revealed the country's backwardness and poverty. In 1994, I witnessed people in Tianjin marveling at the city's first escalator. Over time, I came to appreciate China's diversity, with its fifty-six nationally recognized minorities and hundreds of different languages.

During this project, my days were consumed with a relentless cycle of appointments, factory interviews, and report writing. Each factory visit offered a snapshot of China's industrialization. I developed a keen eye for operational efficiency, gauging it through inventory levels, machinery utilization, and staff engagement. These cursory reviews, combined with interviews, allowed me to piece together a picture of China's evolving automotive component parts manufacturing sector.

The job was far from glamorous. My travels took me to the gritty, industrial heartlands of China, places far removed from the country's burgeoning cosmopolitan centers. Meals with company staff often featured local delicacies that challenged my Western palate—I vividly recall one day when turtle soup was served for both lunch and dinner. In those moments, I longed for the familiar comfort of a pizza or a good hamburger. This culinary culture underscored the vast differences between China and the West, not just in food, but in business practices and societal norms.

The scarcity of Western fast-food chains—there were only twenty-eight McDonald's in all of China at the time—highlighted how far the country still had to go in terms of opening up to global influences. Yet, it also emphasized the unique opportunity I had to witness China's economic transformation from the ground up, unfiltered by Western conveniences or media.

Many of the companies I visited boasted comprehensive facilities for their staff, including dormitories, hospitals, and cafeterias. This self-contained ecosystem hinted at the monumental changes looming on China's horizon—changes that would prove challenging to navigate.

The factories themselves were just the tip of the iceberg. The concept of building a brand, of creating a business with its own identity, was still foreign to most. These Chinese companies were singularly focused on manufacturing products for foreign buyers, content to be cogs in a larger machine.

As I walked through these factories, the future of China unfolded before my eyes. The impending operational transformations would ripple through every aspect of society, touching the lives of hundreds of millions. It was clear that the country stood on the precipice of a new era, one that would redefine not just its economy, but its very identity. The scale of this impending metamorphosis was both exhilarating and daunting, promising to reshape China's role on the global stage in ways few could fully anticipate.

*   *   *

One day during my run up Hong Kong Island, I noticed someone ahead of me. There was never anyone else on the trail that early in the morning. I tried to catch up and saw the person glancing back. He was clearly determined not to let me close the gap. Finally, we reached the top road and he stopped. A moment later, I caught up, breathing deeply. We were, after all, jogging up a mountain.

We introduced ourselves and went our separate ways. Weeks later, we ran into each other again. I asked Michael, "Do you smoke cigars?"

A smile lit up his face.

"How about I bring two Cubans and we meet up?"

He specified a day and time. "Let's meet in my office in the International Finance Centre."

Days later, I made my way to his office. Upon entering, I discovered Michael Pralle was the President of GE. His office was elegantly decorated with Persian carpets. Years later, I would see his place overlooking New York's Central Park, even more grand and elegant.

We smoked cigars right in his office. I had brought two Monte Cristo #2s, which no one could ever complain about. I'd picked them up minutes away at the cigar shop in the Mandarin Oriental. The hotel, ironically, was owned by Jardine Matheson.

I told him what I was doing in China. At the time, that was certainly interesting to business people. Michael shared what GE was looking at, and I remember clearly one thing he said: "Jack Welch told me to get out here and lose some money."

Leave it to Jack Welch to have the most insightful approach to China. He had the proper expectation for 1994 and knew there would be a learning curve. At this time, GE had not entered the country. By 2006, GE would have around $5 billion in revenues from China, and $18 billion in 2022.

It was many years later that Michael and I reconnected. He invested in one of my companies. The last time I saw Michael was at the Ritz Carlton Marina Del Rey in Los Angeles; at dinner with another investor, he challenged me to a push-up contest. We stood up mid-meal and went outside. Despite being ten years older than me, he readily won.

*   *   *

Regarding the $500 million investment fund, years later I discovered that twenty-one investments were made and nineteen of them were unprofitable. There are many reasons for this insanely high failure rate. One of the investments was in a tire manufacturing plant. The manager opened his own factory up the street and even sent defective tires down to the original factory. Another key reason was that all the investments were in the form of joint ventures, which was a big trend in China at the time. You could only invest if you had a local partner.

To this day, I cannot remember a single joint venture in China that, over time, had two happy partners. Failed economic policy and unenlightened investors were the cause of too many failures early on. Leave it to centralized government to set up business parameters they thought were enlightened but ended up being foolish. Jiang Zemin was the president at the time, though this is a consistent affliction that has affected all Chinese presidents.

The tire manufacturing plant example shows how exceptional Chinese are at stealing things. Even today, there are blatant examples of Chinese openly selling products that steal IP or brands. A good example of this is the highly visible market in Shanghai near the Science and Technology Museum. It's full of all sorts of fake goods, including golf clubs, watches, eyeglass frames, and every brand of clothing. The CCP is well aware of this market and lets it operate anyway. The shock is that foreign governments, alone or combined, do not find it useful or possible to tackle such an issue.

During this time, I viewed investments in China with great skepticism. The mandatory "partnerships" often resulted in a lack of control for foreign investors. There was a glaring absence of objective analysis on core business issues; instead of focusing on strategy, business plans, and securing top talent for execution, many investors were seduced by the promise of "access to all of China."

This mindset, which persists even today, is fundamentally flawed. Successful business ventures in China require control and flexibility, not a suspension of reality. The situation has only grown more complex with the recent requirement for private companies to incorporate

Chinese Communist Party cells within their organizational structure.

These experiences taught me the importance of due diligence and maintaining a clear-eyed view of the risks involved in Chinese investments. They also highlighted the need for a nuanced understanding of China's business culture and political landscape. This basically means understanding all the interests of the various stakeholders involved and impacted.

Moving forward, I approached potential Chinese ventures with a more cautious and strategic mindset, always prioritizing control and flexibility over promises of market access. This shift in perspective has proven invaluable in navigating the increasingly challenging business environment in China.

Chinese show no remorse when they steal products or intellectual property. About two years ago, I noticed China was using an electromagnetic catapult for its aircraft carrier. Instinctively skeptical of their ability to develop it independently, I consulted a retired Rear Admiral who confirmed that the Chinese had indeed stolen the technology from the United States.

While the Chinese have shown a propensity for stealing US technology, the extent of such activities remains unclear. The US military's approach to reporting these incidents is complex, balancing public awareness with national security concerns. It's likely that full disclosure of China's technological acquisitions from the American military complex would be deeply concerning to the American public.

From discussions with contacts in China and observations of high-profile incidents, there appears to be a concerning pervasiveness of unethical practices. Examples range from coordinated cheating in international sporting events to widespread academic dishonesty. Such behavior has definitely become normalized within certain segments of Chinese society.

China's rapid adoption of AI will likely exacerbate these issues, potentially amplifying existing unethical practices. Already the use of AI via social media apps by Gen Z is astounding.

In short, the entire China investment fund experience left a bad taste in my mouth. I learned that strategy without responsible execution doesn't make good business sense. Finding the right people to fill an organization chart in China requires looking outside, not just shifting people around internally. This approach, however, goes against historic business practices and the central government's desire to prioritize hiring Chinese. In 1994, the "iron rice bowl" policy— guaranteeing lifetime employment—still cast a long shadow over the business landscape.

* * *

Around this time, I had the distinct honor of meeting Milton Friedman in Hong Kong. He shared an anecdote from his visit to Beijing, where a senior Chinese official had asked him, "Who in America is responsible for the allocation of resources?" The audience laughed. China had become an economic joke, literally.

Friedman viewed Hong Kong as the freest market in the world, enjoying complete economic freedom despite zero political freedom under British colonial rule. He believed free markets would inevitably triumph over communism, famously challenging Phil Donahue: "Tell me, is there some society you know that doesn't run on greed? You think Russia doesn't run on greed? You think China doesn't run on greed?"

During his 1980 visit to China, Friedman's efforts to explain that inflation was a result of government spending fell on deaf ears. Chinese economists, with limited exposure to foreign ideas, simply believed Dr. Friedman didn't understand China, the government, or communism. This default position of dismissing different opinions is something I've encountered many times over the years. I can vividly imagine Dr. Friedman's response to Chinese leaders vocally disagreeing. He'd simply smile and explain the concept from a different angle.

China still struggles with how to make its economy grow faster. The central communist party's control over everything hinders progress. Even free trade zones, when planned, have failed repeatedly in Shanghai. A more logical approach would be to decentralize planning to provinces, allowing them to compete and learn from each other. However, this poses a threat to the CCP's centralized power. Stronger provinces would threaten central control.

When faced with economic challenges, China's leadership often resorts to familiar excuses rather than addressing underlying issues. This reluctance to accept alternative viewpoints extends beyond economics to various aspects of governance and society. When confronted with differing perspectives, the typical response is: "You do not understand China. You do not understand China's culture or history." However, this dismissive attitude overlooks the possibility that outsiders might offer valuable insights precisely because they *can* view China's challenges from a different angle. The irony is that by consistently rejecting external perspectives, China limits its own potential for growth and innovation.

China is even more disagreeable when different opinions are expressed from within its borders. In 2020, Jack Ma insulted China's financial regulators and traditional banks, calling them an "old people's club" and having a "pawnshop mentality." Ma got summoned for questioning and wasn't seen in public for nearly three months. China basically got scared of its tech giants and feared the impact over the CCP's legitimacy that this might have. Even though Jack Ma was correct and could have helped China's financial sector, he was shut down.

In contrast, other countries no longer do anything about stolen IP. They are instead, it seems to me, exhausted. Americans became accustomed to former President Joe Biden not having press conferences. It was only when Trump returned that they were reminded of proper presidential communication and accountability. It strikes me that it is the same between Americans and US politicians— They have dealt with so much China BS over the years they have become insulated from the impact of its insanity.

*   *   *

The Bund in Shanghai beckoned me, its history as a thriving center of foreign capitalist activity, between 1860 and 1930, piquing my interest. Fifty-two European-style buildings, constructed by England, France, Russia, Germany, and the United States, stood as testaments to a bygone era. Once home to the biggest banks, trading companies, hotels, and clubs, even Jardine Matheson had claimed a piece of this architectural marvel.

But as I arrived in 1994, the bustling cosmopolitan area I'd envisioned was a far cry from reality. The stunning buildings had lost their luster, their grandeur dulled by neglect and disrepair. The big financial institutions that once defined the area were nowhere to be seen. In 1949, when Mao Zedong's leadership took control, all the buildings were nationalized and repurposed for government use. The vibrancy had vanished, and by 1994, it had yet to return. China had not just taken the buildings; it had stolen their soul.

It was a late afternoon and the sun was just starting to set. I could imagine being back so many years ago in what was then termed the "Paris of the East". People would be drinking and sharing their adventures. But, I was all alone.

Returning to Hong Kong after my business trip, I felt a wave of relief. The social scene in Hong Kong was vibrant and engaging, much like what I would have expected in Shanghai. Foreigners formed a tight-knit community, an enthusiastic and well-educated group that breathed life into the city. Lang Kwai Fung offered bars that weren't too crowded, and the Captain's Bar in the Mandarin became one of my favorite haunts. The air was thick with cigar smoke, suits were the norm, and strong drinks flowed freely. It was as if Shanghai's spirit had migrated to Hong Kong, bringing with it many wealthy Shanghai families who remain prominent to this day.

During the handover of Hong Kong to China in 1997, I found myself at a party on the Peak. After 156 years of British rule, China was reclaiming a now-prosperous Hong Kong. The evening remains etched in my memory: cloudy skies and a light drizzle set the mood as we watched the solemn British ceremony on TV. China followed this with one of the world's largest fireworks shows, though the clouds obscured the spectacle. It seemed a fitting metaphor for modern China—grand gestures lost in a haze of uncertainty. Despite the historical gravity of the moment, we found joy in drinking and playing Monopoly, an ironic choice given the real-life power shift unfolding around us.

There's a famous phrase in China: *"Shàng yǒu tiāntáng, xià yǒu sūháng"*—"Above there is heaven, below there are Suzhou and Hangzhou." Yet, these once-beautiful places have fallen victim to progress, overrun by industrial parks and teeming with people. The charm and beauty of China have become increasingly elusive, buried under the weight of development and bureaucracy. It's a pattern that repeats itself across the country: wonderful places transformed into something dreary, their essence lost to intense greed, misguided policies, or a sea of visitors. As I stood there in 1997, watching Hong Kong's transition, I couldn't help but wonder if it too would succumb to this fate, its unique character slowly eroded by the tide of change.

CHAPTER 5

1995 Move to Singapore

I n 1995, disenchanted with consulting, I turned my gaze toward Singapore. China's progress in opening up seemed glacial, though in hindsight, the pace was remarkable. Deng Xiaoping's economic reforms had made inroads in southern China, but they had yet to penetrate the country's core. Currency and banking reforms were still on the horizon.

I found myself at United Overseas Bank (UOB) in Singapore, the sole foreigner amidst a sea of local talent. The bank's openness and professionalism impressed me, particularly that of Wee Cho Yaw, the esteemed chairman who has since passed away. My time at UOB offered a unique vantage point to observe Singapore and its remarkable leader, Lee Kuan Yew.

At the official opening of UOB Plaza's twin towers, I witnessed Lee Kuan Yew as an honored guest. Over the years, more than 50,000 Chinese officials have made pilgrimages to Singapore, seeking to unravel the city-state's secret to maintaining order, cleanliness, and contentment while also studying its strict media controls. Deng Xiaoping himself had visited in 1978, setting the stage for a profound exchange of ideas.

Lee Kuan Yew's thirty-three visits to China over twenty-seven years, beginning in 1976, were more than mere diplomatic niceties. Each encounter with Chinese leaders deepened his already profound understanding of the country and its challenges. Lee's influence was pivotal in Deng Xiaoping's decision to open up Shenzhen, which became China's first Special Economic Zone in 1980. This was a watershed moment in China's journey toward foreign investment and market reforms. Inspired by Singapore and urged by Lee, Deng forged ahead with bold market reforms, despite massive resistance. China was at a critical juncture, still reeling from the Great Leap Forward, widespread hunger, and the Cultural Revolution. Deng braved concerns of social unrest, inflation, and bureaucratic inertia to implement these changes.

In his book *One Man's View of the World*, Lee offered a penetrating insight: "For 5,000 years, the Chinese have believed that the country is safe only when the center is strong." He elaborated: "A weak center means confusion and chaos . . ." This encapsulates China's preference for tight control and established order. The fear of *lùan* (chaos) looms large, making the Chinese government wary of experimentation, lest it lead to significant losses, imbalances, or societal upheaval.

The UOB Plaza ceremony, held in an open-air atrium, provided an unexpected metaphor for Singapore's adaptability. As rain began to fall on the Senior Minister, umbrellas quickly appeared to shield him. A colleague quipped, "Good fengshui, bad planning,"—a statement that could equally apply to many of China's subsequent decisions.

Deng Xiaoping's passing in 1997 marked the end of an era. Two years later, the Chinese government's designation of Falun Gong as an "evil cult" revealed its deep-seated fear of movements that might challenge Party authority. China's 2001 entry into the World Trade Organization sparked widespread, but ultimately misplaced, hope in the United States, that China would become more open and a better global citizen.

However, China's track record of honoring agreements has been disappointing. From failing to meet WTO commitments to violating the Sino-British Joint Declaration on Hong Kong, China has repeatedly shown that agreements with it should be approached with deep skepticism, a lesson that seems perpetually unlearned by many.

After five years in Singapore and to seek a change of pace, I returned to Los Angeles. Honestly, I did not want to be one of those people who lost touch with the West. I was also drawn by the burgeoning internet industry and the innovative spirit absent in Asia at the time. I invested with an old friend who had sold his business and developed my own company, backed by presidents from IBM, MasterCard, and GE.

As the internet industry flourished in the United States, China erected its "Great Firewall," a digital manifestation of the Communist Party's desire to control information.

Simultaneously, China embarked on a massive urbanization drive, efficiently administering educational and medical services while aiming to transition from a manufacturing-based economy to a service-oriented one. This was a top-down initiative rather than a market-driven evolution.

This period of rapid change highlighted the stark differences between China's controlled development and the West's more organic growth, setting the stage for the complex global dynamics we see today.

CHAPTER 6

Wenwen, Bingbing, and Manman

Around 2012, it became fashionable for Chinese women to have rhyming nicknames. The Chinese language lends itself to this, often imparting a sense of "cuteness" and making names easier to remember. It's a charming custom that still exists.

I have previously been friends with a Wenwen, a Bingbing, and a Manman. And, I married one of them. Bingbing, however, was quite a character. I met her while living in Pasadena. As our relationship developed, I invited her over for dinner. I still remember her walking down the hallway toward me, in a short skirt, carrying a bottle of Dom Pérignon.

Bingbing owned a home in the Chinese enclave of Yorba Linda, in Greater Los Angeles. On one of my visits, I noticed a dresser-like cabinet full of iPhones and iPads. She would give them away to friends and to people who helped her. Her wealth was significant. She bought only the best brands and, in typical Chinese fashion, cared more about image than quality. Seeing my MacBook Pro, she bought one of her own. Then, she immediately uploaded an illegal version of Microsoft Word that was fifteen years old.

We communicated solely in Mandarin. Despite having lived in America for a few years, she didn't speak a word of English. I was surprised to learn that she had attended business school at Tsinghua University in Beijing, China's most esteemed university. Many of the country's leaders— including Xi Jinping, Hu Jintao, and Zhu Rongji—had attended Tsinghua.

When I asked Bingbing if she knew her university's history, she had no idea. So I explained it to her. During the Boxer Rebellion around 1900, an anti-foreigner, anti-Christian, anti-colonialization event erupted in China. About 250 foreigners were killed, an astounding number at the time. Both peasants and the government became involved. Diplomats, missionaries, soldiers, and others from various Western nations, as well as Japan, took refuge together in Beijing, forming an alliance of eight countries.

The foreign countries sailed ships to China and took over strategic enclaves. Emperor Cixi declared war, despite a significant faction of Chinese wanting reconciliation. The alliance brought in 20,000 troops to China. Outmatched, China eventually agreed to pay an indemnity over the next thirty-nine years to foreign nations, as detailed in the Boxer Protocol. China was humiliated, having taken an overarching defeat to come to terms with the situation.

Theodore Roosevelt felt receiving $30 million from China each year was excessive. In 1909, he obtained Congressional approval to reduce the amount to $10.8 million, on the condition that China use the funds for education. With those funds, China established what is now Tsinghua University. Hearing this, Bingbing was utterly indifferent.

Later, I came to more fully appreciate China's university system. Admission to leading universities is based on test scores, but each university has a quota for students from each of China's twenty-three provinces. This system can make it easier for students from less academically competitive provinces to gain admission. I even heard of one person's father bribing a police commissioner in another province to allow his son to take the Gaokao test there.

I also discovered that Bingbing was the president of her class at Tsinghua University business school, elected by her peers. They had an exclusive WeChat group for all their class members, facilitating easy connections. She regularly reached out to the president of Tsinghua University whenever they needed anything. Everyone in the group was wealthy, from an influential family, or the president of a corporation in China. They were all her friends, and she could call upon them whenever needed.

These were the sort of people who didn't just drive expensive cars; they drove the latest models and sometimes had Bingbing pick them up. When in America, they would fly in everything original to have authentic Beijing hotpot. When we flew to Beijing together, people met us immediately as we stepped off the plane; customs had let them through. They had flowers for us and carried our luggage. Evidently, Bingbing knew the Air China CEO. It was a surreal experience.

Bingbing's wealth and vast collection of brand-name goods came from doing favors for people and her work at a real estate company. Whenever they completed a project, she got to select which condo she wanted at a discounted price. As housing prices rose over time, it was people like Bingbing who benefited greatly. She was an intriguing character, but ultimately not my type.

But, her ability to communicate with all her business school classmates was both impressive and powerful. This seemingly small capability was a distinct advantage over all the other communication platforms I have ever seen.

Later, during a meeting with the University of Chicago Booth School of Business, I mentioned to the head of alumni affairs how China uses WeChat's group function, electing a president for each year, and maintaining contact with their school's president. The UoC alumni affairs person was dismissive, seemingly unaware of the potential such a system might offer for building stronger alumni networks. Or, fearful his role might change.

CHAPTER 7

China 2017

Years passed and I found myself once again drawn to China. Observing from afar until 2017, I was beckoned back to take a look. As my plane descended into Beijing, we plunged through a thick, oppressive cloud that I initially mistook for a dust storm. The reality was far more sinister—a blanket of pollution so pervasive it obscured the city below. I would soon learn that on those rare days when blue skies emerged, people would rush outside, phones in hand, to capture photographic evidence of the anomaly.

In smaller cities, the morning ritual of burning trash persisted, the beautiful smoke dissipating throughout the day. The government's priorities were clear: economic growth trumped environmental concerns, a fact that would make John Kerry's hair stand on end. The populace, it seemed, was more likely to revolt over unemployment than unbreathable air. While smaller coal plants were being shuttered in favor of larger ones, China still consumed more coal than the rest of the world combined.

The taxi ride from the new, sprawling Beijing Airport along a superhighway was a stark reminder of how much had changed. Gone was the charming small highway I remembered; in its place, a network of seven Ring Roads encircled the ever-expanding metropolis.

I had secured an Airbnb in the Diplomatic Residence, a stone's throw from the infamous Sanlitun outdoor shopping mall and a short walk from Tuanjiehu, where I had stayed nearly three decades earlier. The high-ceilinged apartment was a tasteful blend of modern comfort and Chinese curiosities. My unexpected flatmate turned out to be *Le Monde's* chief China correspondent, whose French-accented yet impeccably fluent Mandarin, was a linguistic marvel.

Grateful for the company of someone so knowledgeable about China, I decided to repay his hospitality with a home-cooked meal. One evening, I prepared the finest steak I could procure in China, accompanied by English-style mashed potatoes and stir-fried asparagus. As he devoured the meal with gusto, he regaled me with tales from his years traversing China for work.

Beijing in 2017 was a city transformed. Streets teemed with cars and Didis—China's answer to Uber—driven almost exclusively by migrants from other parts of the country. The economy hummed along at 6.9 percent, buoyed by strong exports and government infrastructure spending. Pollution remained a pressing issue, with half-hearted attempts to curb heavy industry production doing little to clear the air.

I made it my mission to gauge the Chinese sentiment toward the CCP, a task that proved as complex as the country itself. Surprisingly, a majority expressed satisfaction with the party's rule. The country was safe, bellies were full, jobs were plentiful, and homeownership had soared to over 90 percent by 2020. The government's strategy of subsidizing the older generation while leaving the young to the market seemed to be paying off. More nuanced surveys suggested true support for the CCP hovered between 50 and 70 percent—figures that would astonish me.

What struck me most was the disconnect between the average Chinese citizen's perception of their society and the issues that the Western world fixated upon. As I spoke with locals, it became clear that most were either unaware or unconcerned with the increased scrutiny over human rights violations in Xinjiang and Tibet. The government's iron grip on media and online platforms effectively silenced any domestic discussion of these issues. The prevailing attitude I encountered was one of pragmatic indifference.

Most Chinese simply wanted to live their lives undisturbed, focusing on their hometowns and families. The specter of government reprisal—even something as seemingly benign as freezing a WeChat account—was enough to quell any thoughts of dissent. WeChat had become the lifeblood of Chinese society, replacing email and serving as the primary means of communication, payment, and social interaction.

The ubiquity of smartphones in China was striking. For many, especially factory workers enduring grueling conditions, these devices offered a precious escape. The government, in a calculated move, allowed the proliferation of "pirated" entertainment while maintaining strict censorship over sensitive content.

Perhaps the most profound change I witnessed was the coming of age of the one-child policy generation. Contrary to my expectations of encountering a cohort of spoiled youth, I found young adults buckling under the immense pressure to succeed. As the sole focus of their parents' hopes and investments, they carried the weighty responsibility of caring for aging relatives on their shoulders.

The disparities between coastal and inland cities, private companies and government departments, urban and rural areas had become stark. Many individuals left their families behind to pursue opportunities in thriving coastal metropolises, facing the challenges of limited access to social services in their new homes.

As I reflected on these changes, I couldn't help but wonder: How would this complex tapestry of progress and inequality, of satisfaction and underlying tension, shape China's future on the world stage?

CHAPTER 8

Shanghai 2018

Knowing that Beijing was China's political heart, I set my sights on Shanghai, the country's bustling business center. The city's cosmopolitan allure and stunning beauty drew me in again, and I found myself settling into a place in the charming French Concession. It was there, amidst the tree-lined streets and art deco architecture, that a provocative question took root in my mind: If someone could change China, what might they do?

This tantalizing thought sparked the creation of my first thriller, *The Control Center*. The novel, which would become an Amazon bestseller, follows an American in China approached by an Israeli spy with a daring proposition: to infiltrate China's broadcasting center. The writing process ignited a passion within me, eventually leading to a series of four novels.

Before the ink dried on my literary endeavors, another opportunity presented itself. It was 2018, and Chinese investors were eager to deploy their capital in US real estate. However, their vision was narrow, focused solely on the glittering prizes of New York, Los Angeles, and San Francisco. They had yet to grasp the vastness of America's potential beyond these coastal titans.

Collaborating with an old business friend in Los Angeles, a seasoned real estate veteran, I found myself playing tour guide to him and his colleague in Shanghai. Our meetings with potential clients cemented my role as their point person in the city. Among my contacts was a former classmate, now heading real estate investments at one of Shanghai's most prestigious firms.

When a prime opportunity arose—an industrial distribution park in Boise, Idaho—I wasted no time informing her. I forwarded the investment documents and stressed the timeline. Yet, in a move quintessentially Chinese for that era, she dawdled. Three weeks later, when she finally responded, the opportunity had vanished, all funds already raised.

This incident laid bare a fundamental challenge: Chinese investors, even those with prestigious MBAs, often failed to grasp the need for swift action in a competitive market. Their risk aversion and fixation on short payback periods betrayed a lack of sophistication that was hard to fathom. In corporate China, the prevailing mentality still favored inaction over the potential for loss, stifling innovation and growth.

Around this time, I was asked to become a member of the American Chamber of Commerce's Invest USA Committee. I accepted, with the hope I could add value. Companies needed strategic insight and more knowledge about America. Unfortunately, the group was focused on bringing in American companies and holding more of a trade show. China was at a time in its development where it thought it understood America, but it did not, and companies did not have insightful strategies. Chinese investors wanted to do things quickly, rather painstakingly learning and growing from where they were. In Japan, there is a wonderful company that started, over 100 years ago, making canvas bags that were used to carry worker tools. That company evolved and now offers canvas bags in different colors and sizes. They are remarkably successful. This sort of development does not occur in China.

The contrast between this cautious approach and the vibrant nightlife of Shanghai was stark. One evening, I found myself invited to the Roosevelt Club on the Bund by my friend Andrew and his flamboyant boss, Stanley. The venue, I discovered, was steeped in history—originally the Jardine Matheson Building, constructed in 1845, its very walls once housed opium. The Chinese government's attempt to erase this colonial past by renaming it struck me as a futile gesture against the tide of history.

As I approached the building, I paused by the water's edge to take in the Bund's magnificent skyline. Unlike my first visit in 1994, each of the fifty-two buildings now blazed with light, crowned with red Chinese flags. The sight stirred an unsettling mix of awe and disquiet within me.

Inside, Andrew ushered me to their usual private room on the third floor, where Stanley greeted me with a Partagas

D4 Cuban cigar and a glass of Black Label whiskey. As I settled in, savoring the familiar rituals, Stanley announced the arrival of another guest: "Brad, this is Mr. Big!"

Mr. Big, I noted, was ironically the shortest man in the room. His stature spoke of a childhood in China's countryside during leaner times—a product of the Cultural Revolution. As we seated ourselves, I engaged Andrew in conversation, my back slightly turned to our new arrival.

Little did I know, this seemingly innocuous gathering was about to take a tense turn. Suddenly, Mr. Big's voice cut through the room, sharp and accusatory: "Why did you give THAAD to South Korea?" His eyes bored into me, demanding an answer.

He was right to be pissed off, but not at me. The United States gave THAAD to South Korea to help protect it from North Korea, but THAAD could also be used to look deep into China to monitor their missile activity.

I deflected, asking Andrew for more whiskey, but Mr. Big persisted, his tone more aggressive as he asked the question again.

Recognizing the futility of reasoning with a zealot, I finally turned to face him. In measured tones, I delivered a pointed rebuke: "If it weren't for America, you'd be speaking Japanese. You should be more polite."

It was a severe response to a rude question. Truthfully, my answer did have some merit— During the Second Sino-Japanese War in 1937, Japan controlled large sections of China. Without America's participation, Japan could have held on to some of those portions of China.

The room crackled with tension. Stanley, ever the diplomat, hastily raised a toast to friendship and prosperity. As the evening wore on, I found myself locked in conversation with Andrew, a Taiwanese native with a Canadian passport and an unwavering faith in China and its ruling party. His blind optimism stood in stark contrast to the complex realities I had witnessed.

As I bid my farewells, pointedly ignoring Mr. Big's lingering hostility, I opted for a solitary walk back to my flat. The hour-long stroll through Shanghai's neon-drenched streets offered a welcome respite, the city's vibrancy a stark counterpoint to the evening's charged encounters.

In the years that followed, I would often reflect on that night, seeing in it a microcosm of China's contradictions— its ambition and insecurity, its progress and its stubborn adherence to outdated ideologies.

Two years later, I learned of Stanley's ingenious method for circumventing China's capital controls. He established a consulting firm in Hong Kong that provided "services" to his Shanghai real estate business, creating a conduit for funds to flow out of the mainland. It was a tedious yet effective workaround in a system that still struggled to reconcile its global ambitions with its restrictive capital transfer policies.

I pondered these intricate financial maneuvers, wondering how long China could sustain its economic growth while clinging to such restrictive practices. The answer, I suspected, would shape not just China's future, but the global economic landscape for years to come.

*   *   *

The main problem with Shanghai was the Shanghainese themselves. Not terrible people by any means, but they possessed an air of arrogance that set them apart. They seemed to believe they had single-handedly created Shanghai, China's financial center. The truth, of course, was far different. The beautiful city of Shanghai owed its existence to the English, French, Americans, and others who had built it up over the years. Perhaps this misplaced pride was a defense mechanism against the city's colonial past.

Adding insult to injury, Shanghainese cuisine was notoriously sweet and paled in comparison to the rich culinary traditions of the rest of China. Even their distinct dialect was not known for its melodious qualities.

The true treasures of Shanghai are the *wàidìrén*, the outsiders who flocked to the city from all corners of China. These bright, hardworking individuals took enormous risks, leaving their hometowns in search of opportunity. In a city where companies openly favored local hires, the *wàidìrén* had to work twice as hard to secure positions. Yet they persevered, eventually infiltrating all sectors of the economy, including high-tech firms.

Those with less formal education found work in bars and restaurants, cleaning and setting up with unfailing cheer. Most lived frugally, sharing apartments with three or four others to keep costs low. Airport workers often crammed into dormitories with at least six roommates. Their resilience in the face of such challenges was truly inspiring.

* * *

I had the distinct pleasure and honor of dating—and eventually marrying—a *wàidìrén*.

One Friday evening I made my way to Shabbat dinner at Kehilat Shanghai in Jing'an. The community gatherings were always a delight, bringing together an eclectic mix of Jewish expatriates and their Chinese spouses. I'd met legal counsels working entirely in Chinese and executives from Chinese Fortune 500 companies. The group welcomed all with open arms and zero judgment.

That night, however, I was late to the dinner. A café window caught my eye, and within it, two women whose smiles drew me in like a magnet. Abandoning my original plans, I entered the restaurant and approached their table. One of the women, the owner, immediately offered me a seat, while the other woman, Manman, regarded me with shy but genuine interest. In that moment, I felt a surge of gratitude for all the years I'd spent studying Chinese. Little did I know how profoundly this chance encounter would change my life.

After a brief conversation and an exchange of WeChat contacts, I excused myself, my head spinning with possibilities.

* * *

Our first date took place at Haidilao, the famous hotpot chain. I arrived early to secure a table and, in a moment of playful mischief, informed the staff that my guest couldn't use chopsticks. When Manman arrived to find a fork and spoon at her place setting, she shot me a delightfully naughty look that set the tone for the evening.

As we shared the meal, I was struck by Manman's charm and consideration. "Manman" was not a nickname; her name is from the Chinese word that means "romantic" and was given to her at birth. She devoured an impressive amount of vegetables while I focused on the meat, creating a perfect balance. Her background fascinated me—the daughter of garlic farmers from a small town, raised primarily by her grandmother. "I just want a simple life," she said repeatedly, a refreshing sentiment in our complex world.

Yet Manman was far from simple. She worked as a project manager at Tencent, one of China's tech giants, displaying a level of sophistication that belied her limited travels. Her open-mindedness and genuine curiosity about the world around her were captivating. I discovered that Manman did not need a nickname to be cute—she was already adorable.

As I gazed at her across the steaming hotpot, I couldn't help but feel that my life in Shanghai—and perhaps my entire journey in China—had been leading to this moment.

CHAPTER 9

Writing Articles

Around this time, I found myself frequenting the cafés and restaurants of Jing'an, Shanghai's foreigner-friendly enclave. Western eateries, though not abundant, dotted the landscape, offering a taste of home amidst the bustling Chinese metropolis. It was in one such coffee shop that I encountered Peter, a vaguely familiar face that I couldn't quite place. Our friendly chat led to an exchange of WeChat contacts, setting the stage for what would become an intriguing and somewhat unsettling series of events.

Over coffee, Peter revealed his consultancy business, which specialized in economic policy analysis reports. His clients, he explained, were particularly interested in predicting the Trump (first term) administration's policies toward China. When he inquired about my contacts in news organizations and Washington DC, I confirmed that I had some, which was true. Then came the offer: a few thousand dollars to write a report. Intrigued, I agreed.

The task of predicting Trump's actions before they happened was both challenging and oddly exhilarating. I found myself making educated guesses, weaving together snippets of public information and my own analysis. Peter assumed I was tapping into insider knowledge, but in reality, I never spoke to anyone—It was all conjecture and careful reasoning.

As I delved deeper into this world of speculative reporting, a nagging suspicion took root. These reports, I realized, were likely making their way up to the Chinese government. The implications were troubling. If Peter was paying me for these insights, how many others like him were out there, potentially compromising individuals with genuine connections to Washington?

The line between harmless speculation and potential espionage suddenly seemed dangerously thin. Driven by a sense of civic duty, I left a message on the CIA's website, alerting them to this potential risk. I highlighted the particular vulnerability of US-based organizations with Chinese connections, where payments could be discreetly made into Chinese bank accounts, beyond the reach of American oversight.

The situation took a more sinister turn when Peter pressed me to write an article attacking an American politician known for their anti-China stance. This blatant attempt at election interference crossed a line for me. I declined firmly and once again reached out to the CIA, this time explicitly warning them about China's efforts to meddle in American elections.

My brief foray into this shadowy world of international intrigue left me both fascinated and deeply unsettled.

After completing three articles, I decided I'd had enough. The experience had opened my eyes to the very real possibility that well-connected assets in America could be compromised by substantial financial incentives. While not entirely surprising, witnessing firsthand the mechanics of such operations was sobering.

I never heard back from the CIA, nor did I expect to. But as I stepped away from this activity, I couldn't shake the feeling that I had glimpsed something far larger and more complex than I had initially realized. The intersection of international politics, economic speculation, and potential espionage had suddenly become all too real, leaving me with a newfound appreciation for the delicate balance of global affairs.

CHAPTER 10

I Became Hers

After weathering the insecurities that plague nearly every relationship, Manman finally moved in with me. We settled on the infamous Wukang Road in the French Concession, later relocating to the charming Julu Road. I worked from home while Manman juggled a hectic schedule at her office.

One might expect world-class infrastructure in Shanghai, but reality painted a different picture—most residential complexes limped along with internet speeds capped at a paltry 3 Mbps. Outdated routers paired with mismatched modems added to the speed problem. This technological lag spoke volumes about the residents' relative wealth and spending habits. Every yuan was scrutinized, every expenditure carefully weighed. In stark contrast, Manman's workplace boasted lightning-fast internet and unrestricted access to global sites—a rare privilege in China. Such was the dichotomy of Shanghai, then and now.

Our conversations initially flowed almost entirely in Chinese. As time passed, Manman's English improved remarkably. Chinese schools excel at teaching reading and writing in English but falter when it comes to speaking skills. The refrain "There are just too many people in China" echoed throughout the country. Perceived educational inequalities led the government to outlaw private tutoring services in 2021, a decision with far-reaching consequences.

One afternoon as we walked around our neighborhood we noticed a good sized shop under construction. I was pleasantly surprised to see a large sign that said *Peet's Coffee*. I poked my head inside, where they were clearly working. A smiling white face appeared; his name was John and he was a senior roaster checking in on the construction of Peet's first shop in China. To my great pleasure he shared a cup of coffee with us, making us Peet's first customers in China. Now the chain has over 200 hugely successful stores in China, making it another example of living through and participating in the growth of China.

We lived our lives together in Shanghai, each of us from another place. We became intimately familiar with the neighborhood, which would bustle with Chinese tourists on the weekends. To us, Shanghai was home.

*   *   *

One afternoon, a routine trip to the local supermarket became an unexpected lesson in social dynamics. After gathering my essentials, I joined the queue at the only open checkout counter. The monotony of waiting was broken when an elderly woman, visibly frail and limping, made her way to the front of the line. To my dismay, both cashiers dismissed her without hesitation.

Unable to contain my frustration, I called out in Chinese, "For heaven's sake, help the old lady!" My outburst prompted action, and the cashier beckoned the woman over. Glancing behind me, I caught a nod of approval from another customer. His silent agreement only fueled my indignation. "Why didn't you speak up?" I asked him.

As I reached the counter, I confronted the cashier directly. "What becomes of a society that neglects its elders?" The question hung in the air, heavy with implication.

I understood the underlying skepticism. Stories of elderly individuals exploiting their age for preferential treatment had made rounds in the media. But this woman's frailty seemed genuine, her need apparent. The collective inaction I witnessed that day stood in stark contrast to traditional Chinese values of filial piety and respect for the aged. It was a sobering reminder of shifting social norms, made all the more poignant by the fact that it took a foreigner to uphold what was once a cornerstone of Chinese culture.

I exited the market, pausing on the steps and surveyed the bustling intersection before me. At each corner, a handful of people waited for the light to change, their faces illuminated by the glow of smartphone screens. WeChat messages, Bilibili videos, and Xiaohongshu posts consumed their attention, a microcosm of modern China's digital obsession.

Every single person on each corner stood transfixed, eyes glued to their phones. The scene resembled a mass addiction, as if everyone had succumbed to some digital narcotic. Evolving algorithms targeted content and advertisements with frightening precision, molding minds and behaviors in real-time. It was a chilling spectacle, a glimpse into a future where human interaction had been supplanted by virtual engagement. The impact was undeniable, reshaping society with each swipe and click. I couldn't help but wonder if this was the price of progress or the beginning of our downfall.

The irony wasn't lost on me: a nation known for its rich culinary traditions, now filled with people who photograph elaborate meals they can't prepare themselves. They pose in oceans they can't swim in and atop horses they can't ride. When contemporary Chinese speak of their culture, they often refer to a historical legacy rather than lived experience. Beyond the prevalent use of chopsticks, the cultural divide between China and America seems to narrow with each passing year. Yet, China's unique history continues to shape its present in ways both subtle and profound.

*   *   *

As months passed, Manman and I explored every corner of Shanghai, our bond growing stronger with each adventure. Unlike most couples, who sat across from each other in restaurants, we always chose to sit side by side, hands intertwined. Our love was palpable, drawing curious glances from onlookers.

Our favorite pastime was strolling to Jiao Tong University, a mere ten-minute walk from our home. Armed with a picnic and blanket, we'd claim a spot on the edge of the soccer field. As the sun dipped below the horizon, painting the sky in vibrant hues, we'd watch athletes train and families gather. In this bustling metropolis, we had discovered our own slice of tranquility.

During these outings, Manman shared more about her family and upbringing. She reminded me of China's population policy shift in 1984, which allowed couples with a firstborn daughter to try for a son. This twist of fate had gifted Manman a brother, a luxury she wouldn't have had if born earlier. She felt lucky to have a brother who she loved immensely.

At the same time, I witnessed firsthand the rise of a new generation of Chinese women—strong, intelligent, and fiercely independent. The newspapers do not talk about this, but it was truly remarkable. They saw themselves as equals to men, prioritizing careers over traditional homemaking roles. Manman's friends epitomized this shift, representing China's promising future. It wasn't just about taller people due to improved diets or rapid urbanization; it was a fundamental change in how women perceived their place in Chinese society and family structures.

However, China faced a looming demographic crisis. With birth rates plummeting and an aging population, the government grew increasingly concerned. In October 2023, President Xi made a controversial statement, urging women to embrace a "new trend of family." His words, essentially relegating women to domestic roles, revealed how out of touch he was with modern Chinese women. The irony of a leader championing outdated ideals in a rapidly evolving society was not lost on me. As of 2024, American families have adopted 82,674 children from China. Adoptions are no longer permitted.

Another issue compounded the problem: the financial burden of raising a son. Traditionally, men were expected to cover wedding and housing expenses, a responsibility that weighed heavily on families. This fear of having a boy paradoxically led to fewer children overall. President Xi's failure to grasp these shifting societal trends highlighted the growing disconnect between China's leadership and its people.

*   *   *

As our relationship deepened, I realized Manman had captured my heart entirely. Trusting in our bond, I embarked on a trip to Los Angeles to settle some affairs, including vacating my place in Koreatown. Before leaving Shanghai, I asked Manman to visit Cartier for a ring measurement—a hint of what was to come.

Three weeks later, I returned to Shanghai, a small box nestled safely in my briefcase. Manman's radiant smile at the airport reaffirmed what I already knew: I had found my lady. As we rode into the city, the slow pace of China's highways gave us ample time to reconnect. The contrast between China's lightning-fast trains and crawling super-highways never ceased to amuse me. I cherished the slow ride into the city with Manman. Foreigners always marvel at

China's fast trains but never know the superhighway speed limit is 43.5 miles per hour. This fact is highly symbolic and representative of people's rampant misunderstandings of China and the Chinese people. This is not a cultural misunderstanding; it's a factual one.

I'm not one for grand planned gestures, so I kept the ring close, waiting for the right moment. It came unexpectedly the next morning as we lay in bed, our conversation flowing easily. Without fanfare, I retrieved the ring and asked the question that would change our lives. Manman's immediate "Yes!" filled the room with joy.

However, as I went to place the ring on her finger, we encountered an unexpected hiccup. The ring was too large—Cartier had measured the wrong hand. In China, brides typically wear their wedding rings on the right hand, a cultural nuance I had overlooked. Though it needed adjustment, this minor setback couldn't dampen our happiness. As I prepared for one last trip to Los Angeles, I marveled at how life's imperfections often lead to the most perfect moments.

*　*　*

Back in Los Angeles, I mailed off the ring for adjustment. Days later, it returned. It was during this time that COVID-19 emerged in Wuhan. Like the rest of the world, I was mesmerized by the unfolding situation. Manman and I chatted via WeChat video at least twice daily, grappling with the uncertainty that loomed over us.

In China, I belonged to two University of Chicago Booth Business School WeChat groups, each capped at 500 members—a limit imposed to curtail potential coordination and protest. This restriction on communication was typical of China's approach to controlling information flow.

Many marvel at China's manufacturing prowess, but I remain unimpressed. True innovation is rare, and even China's pharmaceutical industry lags behind. They failed to replicate Pfizer's Viagra, and it wasn't until 2017 that they mastered the production of a simple ballpoint pen. China excels at assembly, churning out 38 billion pens annually, but relies heavily on imported manufacturing equipment and expertise. The contrast between China's perceived industrial might and its actual capabilities was stark, a microcosm of the country's broader challenges.

When I suggested in the WeChat group that "The CDC should assist with evaluating the virus," I was promptly expelled from the group by the administrators. As the lone Westerner and voice of reason, my proposal was met with fear rather than consideration. Offline conversations revealed that group members feared government reprisal. This pervasive fear, even in the face of a deadly crisis, exposed a type of endemic paranoia. Simply put, Chinese don't talk about of it but they are scared to death of the CCP.

The government's iron grip has not only stifled free speech but also emasculated Chinese men, who now find themselves dominated by both the Party and a rising generation of strong, intelligent women. Ironically, while the Communist Party historically championed gender equality—with Mao famously declaring, "Women hold up half the sky"—China's current leadership lacks female representation in its top echelons.

My perspective on China's involvement with COVID-19 is straightforward: How could America fund potentially dangerous research by under-qualified researchers? Did the NIH ever properly assess these Chinese labs? And where were the CIA and US Embassy in Beijing when it came to warning about flights from China?

As the pandemic eased in China, I seized the opportunity to book a flight. Waiting at LAX, I indulged in my usual habit of counting newborn Chinese babies—today, I spotted four. This birth tourism, while not malicious, raises questions about citizenship and future obligations.

With approximately fifty weekly round-trip flights between America and China, an estimated 10,200 Chinese citizens participate in this practice annually. Unbeknownst to many, a new generation is growing up with potential access to American passports—and the tax obligations that come with them. This silent demographic shift could have far-reaching implications for both nations.

While I don't view this trend as inherently threatening, the cumulative numbers over years could be significant. Moreover, the influx of about 30,000 military-aged Chinese across the border in 2024 pales in comparison to the broader implications of Chinese influence in America. The real concerns lie elsewhere, in realms yet unexplored by most political observers.

All Chinese in America with familial ties in China constitute a potential security risk. If China wants these individuals to do something, it can threaten them with harassing their relatives in China. This is indeed the sort of thing China would do to force compliance.

CHAPTER 11

COVID in Shanghai

My flight to Shanghai was tense. COVID had recently arrived in America, and everyone on the plane wore masks. Passengers remained motionless except for necessary bathroom trips. No food was served.

Reuniting with Manman consumed my thoughts, making me wish I could accelerate the plane's speed. Finally, we landed and taxied to the gate. I'd lost count of how many times I'd been to this airport, but today was different. COVID had swept through China and was now in the United States. I felt fortunate to be on the flight, somewhat surprised that international travel was still permitted.

The cabin lights remained dimmed after landing. We waited at the gate far longer than usual. An eerie silence pervaded the aircraft, unbroken by any announcements from the flight attendants. Suddenly, a figure in a red hazmat suit entered the plane. He walked deliberately down the aisle to the back, then slowly returned to the front. Three passengers were selected and escorted off the plane. Their fate remained a mystery, the scene reminiscent of a science fiction movie. The surreal nature of the situation sent a chill down my spine, a stark reminder of how quickly the world had changed.

We disembarked slowly, greeted by airport staff entirely clad in hazmat suits, including immigration officials. Handing my passport to a faceless figure felt like arriving on an alien planet. The familiar airport was unrecognizable. Upon exiting immigration, I searched fruitlessly for the taxi line. After fifteen minutes of confusion, I learned I would be quarantined for two weeks. Given two options, I chose the higher-end Shanghai Hotel.

This news came as a shock. I had expected Manman to be waiting, anticipating an intimate taxi ride together. I later discovered that all protocols had changed during our flight. A bus transported us to the hotel, its passengers now prisoners of circumstance. As we carefully disembarked, I spotted Manman. She yearned to approach but was held back by invisible barriers.

The hotel's check-in staff wore hazmat suits, adding to the surreal atmosphere. The Shanghai Hotel, while not ideal for a two-week confinement, could have been worse. I promptly ordered whiskey and snacks online, grateful that package delivery was still allowed. Manman sent two delightful meals daily—hamburgers, tacos, and pizza. I could only imagine how tourists or business travelers without local connections managed.

Despite the circumstances, the Chinese staff at the airport, immigration, on the bus, at the hotel, and those delivering food and packages were unfailingly kind. Their professionalism in the face of this crisis was commendable.

Two weeks later, I was finally released into Manman's embrace. We checked into the Ritz Portman, where my membership status earned us an upgrade to a two-bedroom suite. The hotel's occupancy rate was a mere 13 percent. As Americans began grappling with the harsh realities of COVID, Manman and I found ourselves in an unexpected oasis of comfort.

I couldn't help but feel I had escaped the worst of COVID compared to what others were facing in America. Here I was in China, unexpectedly free—a sensation that seemed to defy the geopolitical tensions between our nations. The irony wasn't lost on me; in fleeing a virus, I had found refuge in a place I once considered restrictive.

* * *

Unlike in America, everything was calm in China. We made plans to get married, which required a trip to Manman's hometown in Shandong to retrieve their family's Hukou booklet—a small ledger recording all official family events.

During this time, I began to understand more deeply how Manman and Chinese express love. Rather than verbal declarations, Manman's love manifested through her actions and glances. She showed care in countless ways. While verbal expressions of love are less common in China, I encouraged Manman to adopt the phrase "I love you." She was a quick study.

It was then that Manman informed me about the Bride Price: $20,000 USD, payable to her father. I was astonished and angered, but not rebellious. Manman offered to reimburse me, but I paid without hesitation. The practice felt archaic, a remnant of ancient China that had somehow persisted into the modern era. After research, I discovered it was still the norm, with prices varying by province. Shandong was among the most expensive. Ironically, this custom now led Chinese families to prefer daughters over sons, and was another disincentive to having children at all.

We proceeded with the formal preparations, having our picture taken—me in a dark blue suit, Manman resplendent in a red *qipao*. She embodied the perfect representation of a beautiful Chinese lady in traditional attire. Against a red background, the photograph was captured, destined for our marriage booklet.

Upon arriving in Shandong, we went directly to the marriage registration office. The process was surprisingly simple: a form to fill, a sentence for me to write in Chinese, and a brief recitation: "I, Brad Good, take you to be my wife, my partner in life, and my one true love." Just like that, we were married in China. Holding the marriage registration booklet, seeing my picture on an official Chinese document, felt surreal—as if I had somehow become part of China's vast history.

Next, we journeyed to Manman's parents' new apartment. Their garlic farm faced abandonment due to underground coal mining, a situation mirrored across

China as progress clashed with tradition. The government's compensation, while fair, often sparked protests—ironically leading to windfall gains for many relocatees as property values soared.

Entering the sixth-floor apartment, I was buoyed by marital bliss, unconcerned about acceptance. Manman's mother initially opposed our union, wary of a foreign son-in-law. Yet over the following days, her father observed our interactions—the small, genuine gestures of care that can't be feigned.

That evening, over an abundant dinner prepared by her father, we shared toasts with a bottle of Kweichow Moutai I had brought. After the meal, Manman's mother looked at us and, speaking in Mandarin for the first time, rather than their local dialect, uttered the profound words: "I take my daughter and give her to you."

Moved by this acceptance, I embraced Manman, declaring, "I take her. She is mine!" This moment encapsulated the cultural shift—Manman leaving her family to join mine, a transition symbolized by the Bride Price.

Manman's father approached, not to toast but to impart wisdom: "Living a simple life is the best. Whenever you have a disagreement, try to understand the other person, and resolve things." We absorbed his advice, appreciating its weight. In that moment, the barriers of language and culture seemed to dissolve, replaced by the universal language of family and love.

Over the next few days, we relaxed with her parents, their local dialect a melodic backdrop to our interactions.

Slowly, I gained an understanding of their speech, joking in Mandarin while never fully mastering their language.

Jining, Manman's hometown and the birthplace of Confucius, held a special fascination for me. The philosopher's wisdom resonated deeply: "I hear and I forget. I see and I remember. I do and I understand." As we embarked on our married life, these words seemed to take on new meaning, a guidepost for the journey ahead.

CHAPTER 12

Dali, Yunnan

While spending time in Shanghai, we knew COVID was still raging around the world. We decided to leave, fearing its potential return. Studying an online map of China, we asked ourselves: *Where is the most beautiful place in the country?*

Without much deliberation, we settled on Dali in Yunnan. We arranged an Airbnb—a two-story haven with a spacious bedroom upstairs and a European-style kitchen and TV area below. A quaint Chinese courtyard completed our private retreat in Old Town Dali.

Dali, home to the Bai people for over 3,000 years, greeted us with a language barrier. Neither Manman nor I spoke the local dialect, but we managed with Mandarin. Perched at an elevation of 6,500 feet, Dali offered a breathtaking canvas of cotton-puff clouds against a deep blue sky. The Old Town nestled beside Erhai, China's seventh-largest lake, while majestic mountains provided enticing hiking trails. This city, once a crucial point on trade routes connecting Tibet, China, Burma, and beyond, had evolved into a tourist haven for Chinese travelers. It had almost a festival feel to it, as if travelers were stopping by, knowing they'd soon continue elsewhere.

With its cobblestone roads and remarkable vistas, Dali invited endless exploration. We decided to stay until we could access Moderna COVID injections in the States—a decision that gave us nine glorious months in this enchanting enclave. While COVID ravaged America, we found solace in Dali's beauty.

During our extended stay, I penned my third novel, drawing inspiration from our surroundings. The book was about the US using AI to help China deal with rebellious internal fractions and the Taiwan situation. Manman, ever ambitious, enrolled in online courses with a US university, bolstering her resume for our eventual move to America.

Simultaneously, we navigated the labyrinthine process of securing Manman's ten-year work visa for America. The application was exhausting, requiring a flight to Guangzhou for an interview. The process left much to be desired, but finally, she received the precious documentation. Ironically, after our marriage, the United States basically granted Manman a passport, while China offered me a mere one-year visa—another example of uneven trade!

Dali surpassed our expectations with its gorgeous scenery and culinary delights. We frequented a sprawling outdoor market, reveling in the variety of fresh produce. One day, in search of potatoes, we discovered five distinct varieties. The market air was thick with the earthy scent of fresh vegetables and the chatter of vendors selling their wares. I selected those resembling russets, perfect companions for the Costco ribeye steaks we'd ordered from Shanghai. We had mastered the art of comfortable living in this far away land.

One afternoon, while enjoying a charming café overlooking a quaint alley, commotion erupted outside. A man's angry voice pierced the tranquility. We rushed to the doorway to find a man brandishing a long stick, poised to strike a young boy of about eight. A small crowd had gathered, watching in silent complicity.

Without hesitation, I intervened, placing myself between the man and child. "You are not permitted to hit children," I stated firmly. The man, likely the boy's father, hesitated before lowering the stick. I returned to the café, my heart pounding.

Manman's reaction was neutral, reflecting the cultural norm of non-interference in parental discipline. While I understood this perspective, I couldn't stand idly by as an emotional adult threatened a child. This incident highlighted a stark cultural difference—I couldn't recall ever witnessing a Chinese person stand up against an injustice. In a country lacking America's diversity, such conformity seemed pervasive.

The traditional Bai people, with their vibrant attire, represented one of China's fifty-six official ethnic groups. They identified as Chinese but maintained a distinct cultural identity, their Buddhist faith evident in their easy smiles. At the market one day, I attempted to purchase three red chilies. The vendor's look suggested I needn't pay for such a small amount, but I insisted.

We learned from expatriates that Dali had once attracted drug enthusiasts, its proximity to Burma, Vietnam, and Laos facilitating easy access. Surprisingly, the Chinese government seemed unconcerned, as users caused little trouble. By the time of our arrival, however, this element had vanished, leaving behind only a few memories held by others.

*   *   *

No one, except for a few older men, swam in Erhai Lake because it was deemed "dangerous." I joined them one day, finding the water delightful and refreshing. In America, the shoreline would be dotted with houses, reminiscent of Lake Tahoe. I envisioned small offices and homes along the lake, attracting businesses from Shanghai. There was ample space. The truth is, the lake isn't dangerous at all; Chinese people avoid it simply because they don't know how to swim. The government, in turn, imposed unnecessary restrictions to "protect" the lake. Heaven forbid the Chinese might have some fun. Wakeboarding is hardly a perilous activity.

We visited Dali University several times. Guards manned all the entrances, so we hiked up the mountain along the university's perimeter. At the summit, we discovered an unlocked door. We'd stroll down past tea plantations on campus, through the academic buildings, and exit where we'd previously been denied entry.

One day near the top, someone asked us in Chinese, "What are you doing here?"

Before Manman could respond, I answered. "I'm a teacher."

Chinese people invariably direct such questions to Manman, not me. Long ago, I learned it was best if I answered; otherwise, a lengthy conversation and more questions would ensue. They assumed I couldn't speak Chinese. But I could, and my answers always satisfied them.

We weren't questioned further. Why was the university closed to non-students? I couldn't fathom. You could easily walk onto Jiao Tong University in Shanghai. I'm suspicious, though. The campus was enormous at 410 acres. I was told there were 18,000 students, but we saw at most 1,000 at any given time, and the campus roads were eerily empty. My guess is that some university bureaucrat simply didn't want tourists wandering around. Unbeknownst to them, they could have developed a world-class wakeboarding program. Think Arizona State. Anyone would love to study at this beautiful university. I checked it out online; they clearly lacked a marketing strategy. Another flawed CCP endeavor. They could have attracted the world's best professors if they'd upped their game.

A handful of foreigners lived in Dali, most owning one of the few foreign restaurants. We befriended them all. Owning a business was financially challenging. Foreign travelers were typically budget-conscious, and the majority of visitors were frugal Chinese tourists. Yet, their spending fueled the local economy. Visitors often rented traditional Chinese attire for photos; Chinese people will do almost anything for a good WeChat moment.

Outside Old Town Dali, closer to Erhai Lake, we discovered a quaint café. It boasted a spacious open-air courtyard and an upstairs area with unobstructed lake views. Breakfast was basic—eggs, toast, and juice—but it offered great coffee, a perfect spot for light work, and stunning vistas. After one visit, I suffered food poisoning—a first in all my China travels.

Weeks later, we returned. Deducing that the culprit had been over-easy eggs, I asked the waitress, "Do you refrigerate your eggs?" Her blank stare said it all. We learned they never refrigerated eggs, deeming them always fresh. Henceforth, in smaller Chinese cities, I promised myself to always order my eggs well-done.

Dali was about a five-hour drive from the real Shangri-La, reputed to offer a unique blend of Tibetan culture and breathtaking scenery. We never felt compelled to visit. If we had, we might never have left. We spent a few nights in nearby towns, all lovely villages with warm inhabitants. One Airbnb we stayed in was on the top of a mountain; our room had unobstructed views. The couple running it were beyond friendly and made us their delightful traditional breakfast both mornings we were there. We also had an unobstructed view of the city below where they burned trash every morning. It was offensive and oddly beautiful at the same time.

It's hard to fathom Dali as part of China, but it is. In Chinese, they say, "*Tiān gāo huángdì yuǎn*"—*The heavens are high and the emperor is far away*. This phrase suggests that places (and individuals) far from central government have more autonomy. That seemed true in Dali. That such a beautiful place could exist without attracting business people or second-home seekers is equally perplexing. People appeared to live free from interference.

The residents of Dali were poor, working hard at manual labor, in restaurants, or shops. They didn't often frequent the lake or stroll the beautiful university campus. Yet, they were content. The smiles on their tanned faces were welcoming and genuine, a testament to the simple joys of life in this hidden paradise. Dali's unique character and relative autonomy contrasted markedly with my experience in other tightly controlled parts of China.

Manman and I discovered Dali together, its newness a shared adventure. Dali's distinct physical beauty and unique inhabitants made it feel like a separate world. As we wandered through the cobblestone alleys, hand in hand, I couldn't help but marvel at how this place, so foreign yet oddly familiar, reflected our relationship.

CHAPTER 13

An Answer for Chinese

Over the years, a few Chinese have candidly asked me what really makes the United States unique, other than free and fair elections. They watch American movies all the time and seem influenced by Hollywood; they cannot really grasp how America is different. Many, in fact, do not think America is that different from China. So, for Chinese in China now reading this book, I would like to share with you my response: "I think all Chinese should know that most developed countries have three branches of government. One is the Legislative, that makes the law. Another is the Executive Branch, which is responsible for carrying out, or executing, the law. Finally, there is the Judicial Branch that interprets and applies the law, and determines whether laws are constitutional.

"The key is that in developed countries, these branches are separate and equal. The president is the head of the Executive Branch and he or she cannot touch the Judicial Branch or influence it. America's Constitution was designed to prevent a crisis between the three branches of government. It was designed to prevent tyranny, and from any one branch getting too much power.

"In China, there is no such clear separation, and this allows for abuse. China faces structural problems, not just people problems. This is something useful for Americans and Chinese to know, but it is only part of the story. The other part is that, in most Western countries, the media is completely free and polices the branches of government. Individuals, most importantly, have real free speech and can bring to light injustices. All of these together, I believe, have enabled Western countries to thrive, compared to countries with other systems."

The above is not a concept that is easy for Chinese to fully grasp. To really appreciate it, the only way is to see it in actions. Even Chinese living in the United States often do not come to appreciate it. I hope more do now.

CHAPTER 14

Los Angeles

Arriving in LA as COVID lingered was unsettling, but we clung to the promise of the new "vaccine." It was a peculiar moment in history. We opted to spend a few days at my friend's place in Agoura Hills, part of Greater Los Angeles. He had been keeping my car, and we planned to embark on a multi-state road trip to scout potential places to live.

I gave Manman a tour of LA, a city I knew intimately. We cruised by my childhood home in Pacific Palisades, then indulged in a late breakfast at Shutters on The Beach. She was captivated by it all. I watched her closely; this was her first venture outside China. She had seen Los Angeles countless times on screen, but being there surpassed her expectations.

The city's vibrant energy and sprawling landscapes seemed to awaken something in Manman, a spark of excitement I hadn't seen before.

A week later, we jetted off to Cabo where my sister's five-bedroom retreat awaited us. My brother flew in from

Arizona, and my sister's two boys joined the gathering. Manman was warmly embraced by everyone. Though hugging isn't customary in Chinese culture, she clearly didn't mind the affectionate welcome.

That evening, slightly overwhelmed, she confided in me: "Your family is different from mine." I knew what she meant—Everyone had open vibrant conversations that night with Manman. They had clearly accepted her as a family member. That was not something she had expected.

Manman discovered the joy of lounging in the pool while gazing out over the ocean, a leisurely indulgence I hadn't seen her enjoy before. We all drank and ate together; one night, Manman made dumplings for everyone. The entire family gathered around, attempting to wrap dumplings themselves. None could compare to hers, each a miniature work of art.

As I watched my family fumble with dumpling wrappers, laughing and bonding with Manman, I realized that love isn't a finite resource—it can be divided and shared.

CHAPTER 15

The Travels

W e spent another few nights in Los Angeles before hitting the road. For Manman, this was a new experience; road trips are rare in China, where people usually travel by train or bus. Under a bright sun, we set out for Las Vegas.

We stayed at the Cosmopolitan Hotel in Vegas and had a balcony view of fountains in front of the Bellagio. It was a great room. We visited the Bellagio to eat, and I watched Manman closely watch the diverse people walking, talking, drinking, and gambling. After dinner, we went to a lobby-level bar in the Cosmopolitan that overlooked many gambling tables. We ordered drinks and enjoyed the view. But my real reason for sitting there was the digital gambling machines in front of us. Manman had never gambled or played blackjack before.

I inserted ten dollars into the machine in front of Manman. It took only a few minutes to explain the game of blackjack to her. She grasped the odds quickly and played like an expert. The ten dollars I put in went down a bit, so I added another ten.

After thirty minutes of play I asked her, "Have you had enough?"

"Yeah," she said, looking at me. You can tell when a Chinese person speaks fluent English. They can smile and say, "Yeah."

"Good. Let's go play for real."

A hint of fear crossed her face.

"Yeah," I said, echoing her word.

I took her hand. "Let's find a good table."

Walking around the casino, we saw many blackjack tables. Some were full. Others were loud with people celebrating, and one had a sad-looking older man. Finally, we found one with a younger couple and joined them. I gave the dealer a thousand dollars and split the chips between Manman and me.

It was then that Manman saw the sign on the table indicating the minimum bet was $50. To calm her, I said, "It's just a game. Don't think of it as money."

This seemed to worry her more, but we started to bet, chatting with the others nearby. Ten minutes passed, and we had won eight out of ten hands. Manman was up over $500, I was up a little bit less. I leaned over. "Let's leave now."

"Can we just go?"

"Yes."

As we stood up the couple next to us looked confused. "You're leaving already?"

In my view, there's nothing worse than losing money you've just won. After walking away, we went straight to the cashier counter.

Manman's face showed more joy than I'd ever seen. She held up the money and receipt, taking pictures of both and a selfie of us to post on Douyin. She briefly wrote about what happened and how much we won.

These are the posts social media users love to see. I'm sure Las Vegas got more value from the post future visitors than the money we won. Getting attention on Chinese social media is both an art and a science. Real happiness almost always wins. Capturing the attention of the Chinese in this day and age is extraordinarily difficult, and Manman had done it. To celebrate our winnings, Manman suggested we go to the hotel's small store to buy some expensive snacks. That was my best night in Vegas, ever.

*   *   *

After visiting Los Angeles, Denver, and Austin, both Manman and I agreed on our preferred destination was Austin. It felt more navigable and youthful.

We found a three-bedroom, two-story condo with high ceilings just north of the University of Texas. With items from my storage, the condo was quickly furnished. Manman set up her office on the second floor. Compared to Los Angeles and Shanghai, Austin was relatively affordable. Our fourth-floor view of trees created a treehouse-like ambiance, markedly superior to anything available in Shanghai.

Manman immediately began her job hunt while I collaborated with my editor on my third book and started writing my fourth. It was a busy period, but we enjoyed our proximity. Weeks passed without Manman finding work, and her frustration began to show. Her strong-willed nature made her reluctant to accept my help, and she relied primarily on LinkedIn rather than networking. The differences in job hunting became apparent, highlighting the contrast between Chinese and American approaches to career advancement.

Eventually, Manman secured a position as a Project Manager at Google—her dream job. She was ecstatic, with a salary exceeding her expectations. Her previous experience at Tencent, a renowned Chinese technology conglomerate, had prepared her well.

Each evening, she'd recount her workday, offering fascinating insights into the differences between working in America and China. From my perspective, she worked thrice as hard in America but earned five times as much—a snapshot of the productivity disparities between the two countries.

I poured my energy into my fourth thriller, all set in China. I took pride in crafting stories that authentically reflected Chinese society and cultural norms, viewing the collective works as a form of art. Writing was both enjoyable and challenging for me.

My investment activity was how I made money at this time. Previously, when Elon Musk smoked pot on Joe Rogan's podcast, I bought a large number of shares the next day after the price massively declined. Upon hearing from Manman that "everyone in China knows Musk," I then sold all my shares. When arriving in Shanghai after COVID had gone through the country, I bought massive amounts of oil on margin, knowing the prices would eventually rebound worldwide.

Getting an MBA at the University of Chicago did not seem of huge use. I often say that I learned investing from a Las Vegas blackjack dealer who advised, "Bet more when you're winning and less when you're losing"—simple yet profound advice.

Throughout this time, I closely monitored US-China relations, acutely aware of the issues outlined in *The China Proclamation*. While researching for my fourth book, I was introduced to a group of remarkable women heading an organization called Lost Voices of Fentanyl. Each member had suffered the unimaginable: the loss of a child to Fentanyl. Their voices rang with clarity and strength, a testament to their resilience in the face of tragedy. The raw pain in their stories cut through the political noise, forcing me to confront the human cost of diplomatic failures.

As I delved deeper into the Fentanyl crisis and America's collective response, a sense of exasperation grew within me. The mothers I spoke with would support any measure that might address the Fentanyl problem. Their desperation was palpable.

Sadly, their efforts—marching, petitioning politicians—had yielded no tangible results. The futility of their struggle against bureaucratic inertia and political indifference was heartbreaking. It became clear that traditional advocacy methods were insufficient in the face of this crisis. Something in America was sorely broken when so many were lost daily without recourse.

As I grappled with these complex issues and put the finishing touches on my novel, Manman immersed herself in her work, blissfully unaware of the storm brewing in my mind. My protagonist, a reflection of my own growing convictions, was spearheading a bold plan to confront China and rectify its injustices. The irony wasn't lost on me—as I crafted a fictional hero's journey to challenge China, my own relationship with the country grew increasingly complicated.

Meanwhile, Manman's dedication to her career served as a poignant reminder of the very changes in Chinese society I was chronicling—the rise of independent, ambitious women pushing against traditional expectations. Her success was a testament to the complex, evolving nature of modern China.

CHAPTER 16

Book Launch 2022

In October 2022, I successfully launched four thrillers, one month apart. *The Control Center* became an Amazon bestseller. I started appearing on China-in-Focus, speaking about China on TV.

It was then that Manman had a realization, followed swiftly by one of my own: My negative commentary about the Chinese Communist Party could have far-reaching consequences for her and her parents. She was deeply concerned, and I found myself uncertain, compelled to investigate the matter further.

The gravity of our situation slowly dawned on me, a chilling reminder of the CCP's far-reaching influence. Speaking with knowledgeable contacts, it became clear that there was a significant chance her parents would face harassment because of my actions. It wasn't uncommon in such situations for parents to lose their jobs or pensions, their livelihoods hanging by a thread.

The likelihood of repercussions for Manman's parents, stemming from my words and writings, was alarmingly high. A heaviness settled in my chest as I grappled with the unintended consequences of my actions.

As the full implications of our predicament sank in, I found myself torn between my principles and the potential harm I could bring to Manman's family. The thought of her parents suffering for my outspokenness filled me with a mixture of guilt and defiance. How could I reconcile my desire to speak truth to power with the very real threat to those I cared about? The answer, I realized, would not come easily.

My books were more than just time investments; they had become a part of me, something I was loathe to give up. Manman expressed her feelings through anger toward me. Soon, it became clear the situation was untenable. With a mix of sadness and anger, we both retained attorneys. The process turned contentious as Manman struggled to understand she had no claim to money I'd earned before our marriage or to investments made with that money.

In Texas, mediation is required before seeing a judge. In my attorney's office, we spoke with the mediator, who shuttled between Zoom rooms to communicate with Manman and her lawyer. Eventually, we reached a reasonable resolution. At the call's end, the mediator advised me to "spend the night someplace else" due to Manman's distress.

That night, like all nights before, we slept in each other's arms.

She moved out. Two years after buying it, I sold our beautiful condo. The night before my departure for Taiwan, we stayed together at a hotel near the Austin airport. With two pieces of luggage, leaving behind nearly everything I owned, including the one I loved, I left.

I never suspected in my wildest dreams that I would become a victim of the Chinese Communist Party. But I refuse to let the CCP live rent-free in my mind. Just over a year later, writing this in Taipei, I ask myself, "What happened? How could that happen? Where is my Manman?"

*   *   *

What happened to me serves as the driving force behind this book. However, it's crucial to understand that these events didn't shape my views; rather, they reinforced convictions I already held. My fourth thriller, in fact, articulates these perspectives that predated my personal ordeal. The irony wasn't lost on me—fiction had become a harbinger of my reality. It sucks.

I feel this distinction is essential for readers to comprehend my motivation. My experiences with Manman and the far-reaching consequences of my outspokenness didn't create my stance; they merely crystallized it, transforming existing concerns into painfully tangible realities on paper.

As I sit here in Taipei, penning these words, I'm acutely aware of the weight they carry. Each sentence is not just an expression of thought, but a testament to the price of speaking out. The pages before me are no longer just paper and ink, but a battleground where personal loss and unwavering conviction meet. Yet, despite the cost— or perhaps because of it—I find myself more determined than ever to share this story, which is not just mine, but the larger narrative of those silenced by fear and oppression.

This book, then, is *not* born from a desire for retribution or a need to justify my actions. It's a continuation of a journey I embarked upon long before the CCP cast its shadow over my personal life. It's a commitment to the truth I've always sought to unveil, not tempered by the harsh realities of its consequences.

In the end, this isn't just my story. It's a mirror reflecting the countless untold tales of those who've faced similar choices, similar losses. And in sharing it, I hope to illuminate the path for others who find themselves at the crossroads of conscience and comfort.

SECTION II:
THE TAIWAN SITUATION

I n Taipei I live around the corner from the Chiang Kai-shek Memorial. Chiang faced numerous challenges fighting the Japanese invasion and later against the Chinese communists. After Japan's defeat, Chiang received vital assistance from Truman to combat Mao's communists. However, the Truman administration soon realized that Chiang's government was thoroughly corrupt, and further aid would be wasted or pocketed by Chiang himself. In 1950, Truman announced the end of assistance, stating, "They're thieves, every damn one of them," referring to the Nationalist leadership. "They stole $750 million out of the American treasury."

This decision contributed to Chiang's retreat to Taiwan and his defeat on Mainland China. Had it not been for Chiang's corruption, the course of Chinese history might have been dramatically altered.

Ironically, America remains entangled in the Taiwan-China relationship. It's crucial to reiterate that issues like Chinese surveillance balloons over America, Uighur human rights concerns, the Fentanyl crisis, and even Taiwan are symptoms of a larger problem: the Chinese Communist Party's leadership. Rather than focusing solely on arming and training Taiwan, the United States should develop a comprehensive strategy to address the root cause. America's lack of such a strategy has created an environment conducive to China's transgressions and led to the Taiwan situation.

China will eventually gain control over Taiwan, and America must understand the implications. *While China's military position isn't yet overwhelming, it is in America's best interest to now help Taiwan negotiate favorable terms with China.* There are a number of arguments justifying this recommendation:

First, delaying will only weaken Taiwan's bargaining power. China will resort to military invasion if necessary, and the US Seventh Fleet cannot indefinitely defend against China's growing forces. Lee Kuan Yew aptly stated in his book, *One Man's View of The World*: "The future of Taiwan is not determined by the wishes of the people of Taiwan. It is determined by the reality of the power equation between Taiwan and China." I wholeheartedly agree. China's military buildup, the most substantial since World War II, saw defense spending reach $230 billion in 2022. While not yet capable of fulfilling all its geopolitical ambitions, China is rapidly developing a formidable presence in the Indo-Pacific Region.

Taiwan is a core issue for China but peripheral for America. China has always viewed Taiwan as part of its territory. Taiwan's 1989 political transformation toward democracy doesn't alter China's historical perspective. Having lived across China, Taiwan appears to fit in almost perfectly from a cultural and social perspective as another Chinese province. China's desire for unification is understandable, given the linguistic and cultural similarities.

Second, China's approach is not purely military. It employs PR firms to disseminate propaganda within Taiwan, recruits top Taiwanese talent, uses trade as leverage, and even bribes influencers. These non-military methods are likely to intensify, and Taiwan struggles to combat them effectively.

Third, America's current strategy of aiding Taiwan defensively and selling weapons like THAAD is insufficient. Defensive capabilities alone will not deter China. Without the ability and commitment to offensively strike key areas in China, there's no genuine deterrent. Taiwan's limited missile range of 1,200 km poses little threat and would likely provoke disproportionate retaliation.

Fourth, Some of the very real issues are not discussed in the press or within think tanks. The Taiwanese are, in a sense, like a person in jail waiting to be sentenced to prison by a judge. Right now they do not know when China will invade. It's like they do not know how long their prison sentence will be and they are genuinely scared. This uncertainty is psychologically distressing. It looms over the people in Taiwan—I see it daily. Such emotional stress can be so debilitating that it impacts people physically. People in jail often prematurely rush to settle a deal just to have certainty. They regularly settle for less than what they could have gotten. The Taiwanese are not in control of their own fate and they see evidence daily justifying this psychological toll whenever they read the news or pass by one Taiwan's 105,000 air raid shelters.

Fifth, has the risk of a China invasion resulted in decreased investment, less real estate development, and the exodus of wealthy Taiwanese? I think so, but this is not easy to quantify. The cost of factory space in Taiwan is roughly equal to that in Shanghai. Shanghai has a moderately lower wage and cost of living. So, why has Shanghai boomed so incredibly while Taiwan has a democratic government and better legal system? I attribute it to invasion risk. Visual support of this can be seen from a real estate standpoint. There are very few new modern developments to replace 1970's outdated structures. The point I'm making is that should China reunify with Taiwan, then the island would likely boom since the China risk factor is no longer there. This is an argument in support of reunification.

Sixth, a Taiwanese acquaintance wonderfully captured a local sentiment: "If only China would try to woo us like a man does a lady." Many view a potential Chinese takeover through this lens. Uber drivers seem indifferent, saying, "What will happen will happen." Students even speculate about increased job opportunities under Chinese rule. Another elder man mentioned, "Taiwan is like a wife that has been beaten one too many times, and has decided to leave." You get different perspectives when you talk to the younger generation versus older or middle-aged. You cannot simply walk around and speak with people and have a statistically robust picture of what Taiwanese think and want.

The only objective gauge of populace sentiment are votes. Taiwan's democracy complicates matters. In 2024, the pro-China reunification candidate secured an astounding 33.5 percent of the popular vote, with over 46 percent of legislative seats going to pro-China candidates. This political landscape could shift further, potentially resulting in a reunification-friendly government. Many Taiwanese, having never visited Mainland China, underestimate the implications of reunification.

Finally, the Chinese in China view Taiwan's "reunification" as inevitable. China is not in a rush and may wait until after Trump's second term. But, they view Taiwanese as Chinese, just like Shanghainese are Chinese, which severely hinders any appetite China might have for a bloody invasion. However, China's determination to reclaim Taiwan is unwavering, as evidenced by its handling of Hong Kong.

The key priority for America to assist Taiwan is to develop and execute a China strategy. While America should help delay reunification with China, this is a futile exercise in the long run, unless decisive action is taken to stop piecemeal policies and actions. It is this exact type of mentality that has led to this current awful situation, that could have otherwise been avoided.

SECTION III: AMERICA'S PAST EFFORTS

f you read The China Proclamation below, you'll find a factual account of China's transgressions against its own people and the world. This is indisputable. How has the United States responded? Rather than detail a complete analysis, I'll comment on the key actions of the first Trump administration, the Biden administration, Congress, the media, the military, China experts, and pundits. Then, I'll highlight the flaws in their logic. Except for Trump, their efforts have yielded no positive results. All parties, save Trump, have mischaracterized China and failed to create or articulate an effective strategy.

FIRST TRUMP ADMINISTRATION

The First Trump Administration marked a significant shift in dealing with China. After Trump's term, China sanctioned twenty-eight top administration officials and immediate family members, freezing their assets and barring them from travel to China. This was in response to the administration's vocal stance on Hong Kong and the National Security Law, as well as strengthened relations with Taiwan and sanctions imposed for human rights violations.

Trump's approach was insightful and educational. In July 2020, he had Mike Pompeo, Robert O'Brien, and Bill Barr deliver speeches explaining China's human rights violations and threat to national security. Trump grasped China's nature. His administration's actions were a stark departure from the diplomatic dance of previous years, cutting through the fog of polite rhetoric to expose the harsh realities of the CCP's governance and underlying philosophy.

Former national security advisor, Robert O'Brien, described US policy toward China as a massive failure: "This miscalculation is the greatest failure of American foreign policy since the 1930s." As Pompeo said at his Nixon Presidential Library speech, "We must induce China to change in more creative ways because Beijing's actions threaten our people and our prosperity."

China's actions threaten its own people too, and America has not yet come up with a "creative way" to deal with CCP leadership. Both gentlemen knew of the threat but did not have a way of approaching it. They knew of the problem but not the solution. This has been case for too long and is the fundamental reason for this book.

There are four important insights worth noting about China sanctioning twenty-eight Trump officials: *First*, China deeply cares about its image domestically and globally. Trump's team was perceived to have tarnished this image, exposing a vulnerability.

Second, China's measured response, waiting until the administration's end, revealed fear of retaliation.

Third, China's anger at the public exposure of its actions was palpable. The insular CCP leadership failed to grasp how their sanctions would be perceived globally. They were not even noticed by the new administration, and ignored by the world.

Fourth, China projected what it viewed as serious consequences, revealing its disconnect with international realities. Those twenty-eight officials sanctioned by China couldn't have cared less.

The key takeaway: China is hypersensitive to its global image yet struggles to manage its reactions effectively.

Trump's trade agreement resulted in $124 billion in additional purchases, though falling short of the agreed target. He banned Huawei and ZTE, restricted China's access to American technology, and took concrete steps to protect America. But, he did not go far enough—at least not yet.

BIDEN ADMINISTRATION

The Biden administration consistently framed the China relationship as "complex," mirroring rhetoric used in other contentious debates. Biden's indecisiveness and lack of understanding the bigger picture was evident in the handling of the Chinese surveillance balloon incident. Sanctioning balloon parts manufacturers was exactly what not to do in order to deal with the China problem.

Biden's mistaken labeling of Xi as a "dictator" inadvertently touched a nerve. While a diplomatic faux pas, it challenged Xi's carefully cultivated image as a strong yet benevolent leader. This incident underscored the importance of language in international diplomacy. The gaffe, while unintentional, revealed Xi's vulnerability.

Surprisingly, Biden maintained many of Trump's policies, including tariffs. His "Strategic Competition" approach focused on domestic investment and international alliances to counter China's rise. However, this piecemeal strategy failed to address critical issues comprehensively, and had zero impact on the Fentanyl crisis.

The administration's frequent diplomatic missions to Beijing—including visits by John Kerry, Janet Yellen, Tony Blinken, Gina Raimondo, John Podesta, and Jake Sullivan—projected an image of desperation. These visits, while intended to show engagement, signaled weakness.

Americans' compulsion to communicate often translates as nervousness or weakness in negotiations with Chinese officials. The parade of officials to Beijing conveyed dependence on China. A more effective approach would be to act independently, waiting for China to initiate contact.

At the height of Biden's engagement with China, Secretary of Treasury Janet Yellen bowed to Chinese officials during her Beijing visit. Yellen had confused Chinese with Japanese etiquette. Ironically, this faux pax not only embarrassed her in China, but the Japanese in Japan also noticed the bowing was too repeated and rapid to have been meant for them. This is emblematic of Biden's approach to China.

CONGRESS

Congressman Mike Gallagher (R-Wisconsin), the previous Chairman of the House Select Committee on the Chinese Communist Party, led a failed bipartisan experiment to address symptoms of the China problem. Their approach lacked a cohesive strategy, focusing instead on adopting legislative programs targeting issues like Taiwan, Xinjiang, and human rights advocacy.

During "China Week" in the House, the BIOSECURE Act passed, prohibiting federal funding for services provided by biotechnology companies deemed concerning, particularly those based in China. One of Mike Gallagher and Matt Pottinger's most significant legislative successes was banning TikTok on government devices—a narrow approach that missed the broader strategic implications. This piecemeal approach to policy-making reflects a fundamental misunderstanding of the challenge posed by China.

Once again, there is an important Chinese phrase: *Zhìbiǎo bùzhìběn*—treating the symptoms but not the root cause. This aptly describes Gallagher and Pottinger's approach. America can no longer afford to get sidetracked with such thinking. A comprehensive strategy is necessary; anything less is negligent and unacceptable.

UNITED STATES MILITARY

The US Military, while recognizing China's threat, has failed to push for a comprehensive strategy. Rear Admiral Mike Studeman's insights, while valuable in defining the threat, overlooked the necessity of a well-defined strategy. As Sir Lawrence Freedom noted in a Joint Chiefs of Staff document: "Without a strategy, facing up to any problem or striving for any objective would be considered negligent." It's fascinating how this quote was recently removed from the document, raising questions about the military's current strategic thinking.

Really, the US military should work with the administration and take actions to support an overarching strategy. I did not hear anyone in the military say this.

CIA/NSA

The CIA has an unofficial motto regarding the officers they prefer to recruit: "We're looking for people who have a PhD and can win a bar fight." When it comes to China, this motto requires a nuanced modification: "We're looking for individuals who appear Chinese, speak native Mandarin, possess a PhD, and can win a brawl in a Shanghai nightclub."

Yet, the reality of the CIA's recruitment efforts paints a starkly different picture. The agency now finds itself recruiting white American teenagers who have barely dipped their toes into the vast ocean of Mandarin studies.

The CIA/NSA dangle the promise of a fully funded college education to entice these youths to persevere in their Chinese language studies. This strategy is misguided and reeks of desperation. I'd instead focus recruitment efforts in Taiwan and Hong Kong.

Recent events have seen China successfully pilfer US military blueprints for both the electromagnetic catapult used on aircraft carriers and the cutting-edge F-35 fighter jet. While the CIA/NSA may boast formidable cyber weaponry, they've fallen woefully short in safeguarding America from China's targeted assaults on US infrastructure, government agencies, officials, and telecommunications networks. Many analysts view these incursions as mere dress rehearsals for a more devastating attack. The digital battlefield has become a new frontline, and America finds itself outflanked.

The 2020 SolarWinds hack, attributed to Russian state actors but potentially involving Chinese collaboration, serves as a chilling example of the vulnerabilities in America's cyber defenses. Perhaps most alarmingly, the CIA's presence in China failed to sound the alarm when the virus from Wuhan silently boarded planes bound for American soil, a lapse that would have far-reaching consequences.

When operating effectively within a country, the CIA can excel at its primary duty of gathering intelligence. Should tensions escalate to armed conflict, this intelligence may prove vital. However, raw information alone often proves insufficient when grappling with China's complex global ambitions. What's needed now is a cadre of operatives capable of not just collecting data, but interpreting it—individuals who can report on what they hear, what they've learned based upon deep experience, predict future developments, and analyze the potential ramifications.

This is the true nature of today's clandestine warfare. Ultimately, strategies are needed. The CIA should be in a position to offer viable options to policymakers, or at least provide suitable input information to enable their formulation. Despite lacking confidential operational information, I remain exceedingly skeptical of the agency's current capacity to fulfill this crucial role.

US NEWS MEDIA & PUNDITS

American news media's issue-focused approach, while attracting views, fails to address the need for a comprehensive China strategy. The media's tendency to sensationalize isolated incidents, like the Chinese balloon incident, distracts from the broader strategic picture. While this approach makes sense for ratings, it does a disservice to public understanding.

The media is complicit in platforming self-proclaimed China experts who have been consistently wrong in their assessments. Gordon Chang, author of *The Coming Collapse of China* (2001), exemplifies this problem. Despite his disproven predictions, Chang continues to be featured, offering alarmist views without proposing viable strategies.

This echo chamber of failed predictions and sensationalism not only misleads the public but also undermines genuine efforts to formulate effective policies.

Media interviewers often fail to challenge guests on strategic solutions, allowing vague responses like "We need to get tough on China" to go unchallenged. This journalistic malpractice perpetuates a cycle of problem identification without solution-seeking.

The media's handling of the Fentanyl crisis is particularly unfortunate. Journalists express sympathy to victims' families but fail to demand specific actions or admit the historical scale of the unaddressed problem. This superficial coverage does little to drive meaningful policy changes or public understanding.

*   *   *

In short, there has been a lack of focus on developing a strategy toward China. We've learned, often through costly missteps, what approaches are ineffective. Like a skilled poker player who's accidentally shown their hand, China's reactions to various US policies have exposed chinks in their carefully constructed armor.

As we stand at this critical juncture in Sino-American relations, it's clear that the old playbook is obsolete. The challenge before us is not just to address individual issues but to craft a comprehensive strategy that acknowledges China's complexities while maintaining American values and morality. The stakes are too high for continued improvisation; we need a grander strategy that turns these hard-won lessons into actionable policy.

It is also vital to appreciate the Chinese mentality of permanence—Their government is permanent. They are not only proud of their history, but know that it is the long game that is important. Placating Westerners in the short term is fine since a time will come when further progress on longer term objectives can be made. Americans like to talk about everything and come to an agreement. "Let's talk about what we have in common" is a phrase commonly used by US administration officials. China is vastly more action-oriented. In a dictatorship, the appearance of strength is highly valued.

There are all sorts of traditional Chinese values. But, the country has changed so much over the years and its values have eroded. It is my hope that Western nations understand the issues outlined below and then act decisively, with fierce determination. Anything short of this will simply yield the status quo.

SECTION IV:
STRATEGIC FOUNDATION

"**X**inwen Lianbo," China's evening news broadcast since 1978, remains a national ritual and the party's mouthpiece. It shapes citizens' impressions of their government and the United States by showcasing busy leaders, rapid national development, and global chaos. While mobile news has gained prominence, this medium has already molded the minds of older generations. The subtle messaging in these broadcasts reveals much about China's self-perception and worldview.

China's narrative is built on two pillars: the "Century of Humiliation" and a deep-seated pride in its 5,000-year history. This duality shapes Chinese attitudes, policies, and global perspectives. The result is a complex national psyche, combining insecurity and determination.

The "Century of Humiliation" instilled a fear of foreign influence and a resolve to prevent similar occur- rences. Simultaneously, China's long history fosters pride in cultural achievements, national unity, and resilience. These forces have cultivated a heightened sense of shared struggle and unity.

China's political and foreign policy aims to reclaim its position as a global power. This drive manifests in the desire to secure Taiwan and resolve territorial disputes. A deep-seated skepticism of foreign influences underlies these ambitions, guiding China's economic and military development.

The pendulum of China's self-perception swings between past greatness and historical humiliation, creating a nation perpetually on guard. This historical context explains China's reluctance to negotiate, even when logically beneficial. The fear of compromise stems from the perception that it could lead to a repeat of past humiliations.

Understanding these realities is crucial when developing strategies to engage with China. The Chinese people and CCP leadership can endure significant hardship when faced with potentially humiliating situations, particularly those instigated by foreign influences.

Any effective strategy must leverage these insights. It should create "fear" within the CCP of reliving past humiliations or losing legitimacy while simultaneously offering "hope" to the Chinese people. Only by pressing these two emotions can China be guided toward proper behavior within its borders and the global community.

The West's failure to grasp this psychological landscape has led to repeated diplomatic missteps. America cannot wait for China's historic national dignity to be restored naturally; the only path forward is to challenge that historic dignity, prompting China's leadership to evolve or change. The key question is how the West can do this while maintaining moral rectitude.

President Xi Jinping, one of China's most forceful leaders in decades, has been ruthless in purging political adversaries. Loyalty is pledged not to the party but to Xi himself, solidifying his position as a dictator with control over all aspects of governance.

A Chinese military strategist once said: "Appear weak when you are strong, and strong when you are weak,"—and this applies to Xi's leadership. While he projects strength, discontent simmers beneath the surface in various aspects of Chinese society. Protests, economic concerns, fear among the wealthy, and internal criticisms of Xi's political ideology reveal vulnerabilities. Every dictator has strengths, weaknesses, and vulnerabilities—understanding and leveraging these will be the foundation of any successful strategy.

Any approach to China must also appreciate not just its long history and what it believes is its culture, but the associated emphasis on strategic patience and long-term thinking. This traditional perspective can and should be challenged. Efforts should be made to move China from strategic patience to strategic urgency. Outside shocks can do this. And, measures to create internal pressures can challenge the notion that China can always afford to wait.

*   *   *

In 2024, the Pew Research Center conducted a thirty-five country survey in major countries, assessing people's views of China. Eighty-one percent of Americans view China unfavorably. In Japan, the rate is 87 percent; 67 percent in the UK; and 85 percent in Australia.

This common belief has resulted in nothing. All efforts to change things by any one country have failed. Any approach to China must appreciate the need for a global front as an imperative of any strategy adopted. Otherwise, countries will fear repercussions. The world stands united in its wariness, yet paralyzed in its response.

It is important to clarify that China is not the problem, nor is the Communist Party. Rather, it is *the senior leadership of the Communist Party* who have authority and responsibility. *Their behavior must change.* Once it is agreed upon that this is the strategic objective, the question is then *how* it will be achieved.

The initiatives below help further define the strategic objective and constitute an approach to deal with the senior China leadership problem. This approach has taken into consideration the nature of China's transgressions, the psychological perspective of CCP leadership and citizens, the international community, and what are just actions to achieve desired results. After all, the end *never* justifies the means.

What follows is not merely a list of policy recommendations, but more of a blueprint to shift global dynamics. Each initiative is a carefully placed domino, designed to reinforce the previous one, ultimately achieving the strategic objective.

The order of the initiatives is critical. People often speak in generalities that cannot be clearly understood or put into action. That must be avoided and will help foster a discussion for continually improving a plan during its execution. Close coordination between stakeholders within and amongst Western governments must occur to prepare, gain input, and anticipate China's responses.

Three critical factors serve as the basis for the initiatives. *One*, they are not meant to, and should not, cause military conflict. *Two*, they should not harm the Chinese people. And, *three*, they are not meant to cause economic damage. Economic repercussions are inevitable; it is important that Chinese citizens appreciate they come from China leadership, and not Western nations. Ideally, our strategic objective can be achieved with minimal economic damage and without economically harming Chinese citizens.

The initiatives should unfold more like a play rather than a military exercise. If initiatives are announced at weekly press briefings, both the press and China will come to anticipate the initiative on its due date. Initiatives should not be revealed beforehand. Uncertainty can then breed fear. And that is something we definitely want . . . to instill fear . . . fear of being humiliated.

In 1973 Nixon said, "My rule in international affairs is: Do unto others as they would do unto you." Kissinger then added, "Plus 10 percent." That's all we are doing with these initiatives but, since we haven't been acting correctly for many years, we need to catch up and add a little bit of interest. We are not taking the Jimmy Hoffa approach which would be: "Do unto others before they do unto you, and make sure they can't retaliate."

SECTION V:
THE CHINA BLUEPRINT

INITIATIVE 1: WRITE & SIGN
THE CHINA PROCLAMATION

The Chinese have no appreciation for the depth of their government's transgressions. And, frankly, they do not care. As mentioned above, Bingbing did not even care to learn about the remarkable genesis of her graduate university. I was kicked out of a WeChat group after suggesting China let in the CDC to check out COVID. This demonstrates a deep fear Chinese have of being embarrassed by any Western perspective or involvement.

Congress, media, pundits and others are issue-focused, not overall-problem-focused. Every person and country must start viewing China as a problem overall, discuss it, and be unified that concrete actions must be taken.

* * *

To do this President Donald Trump should select someone in his administration to factually detail China's transgressions, collaborating with equivalent individuals from both England and France. I call this The China Proclamation. (See Section VII for an example of this document.) From

Day 1, all countries must agree that such internal (within China) and external offenses should be documented. This alliance would serve as the cornerstone for a unified Western response, lending credibility and weight to the document's contents.

The individual selected from the U. should ideally be the frontman for ongoing press briefings. Preferably, President Trump should defer to that individual when speaking with the press, through closely coordinating behind the scenes. This will help remove the president for the hassle of incessant questions regarding China.

The document should be formal and similar in structure to The Declaration of Independence. It should have an ending that places the blame for the offenses not on China, but on *senior CCP leaders*. Ideally, the end of the document will simply state: *The senior leadership of China's Communist Party must change their behavior.* The document is meant to be factual and noncontroversial. It should be signed by all Western leaders.

The importance of this document cannot be overstated. It specifically mentions a strategy at the end: *The leadership of China must change their behavior.* Detailing such a proclamation should not take more than a few months. Since all Western countries need to sign, other countries should be sought out for their input, as their involvement will help ensure their buy-in.

When it comes to Germany, resistance can be expected since its neighbor is Russia, China's BFF. I would propose that if Germany does not want to help Western civilization with China, America should not help Germany in Europe. In other words, consideration could be given to relocating the 50,000 American troops currently stationed in Germany. A portion can be moved to Japan. India will sign at the threat of product sanctions, if necessary.

The China Proclamation will humiliate China. It delegitimizes China's leadership and will remind both the leadership and citizens of its past. Ultimately, China will face global humiliation, and in this case, it means it's the start of incentives to stop doing horrible stuff to its own people and others. The proclamation is *not* a legal document, but it must be factual with sources. It is not propaganda. It is not Western countries trying to meddle in China. It will stand throughout history.

In short, this document will strike at the very heart of China's carefully constructed narrative. It will be a mirror held up to the CCP leadership, reflecting back the ugly truths they've tried so hard to conceal. For Chinese citizens, it could be a watershed moment, a crack in the facade of propaganda that has shaped their worldview for decades.

When everyone works together, China becomes vastly more vulnerable. For example, worldwide citizens could now be urged to avoid buying all products made in China. Such a threat *could* be issued. But such threats, even if large scale, will likely not have an impact, given China's resolve. Just like kicking China out of the WTO, this would *not* help achieve our strategic objective and is not along the lines of what is outlined above under Strategic Foundation.

Execution is key. The document should be translated into Chinese and made available everywhere, including Taiwan. There should be White House press briefings. Chinese subtitles should be used and video clips made for posting on all global social media platforms. There are about 275,000 Chinese students in America. Clips should be suitable for Chinese to share globally with family and friends in China. Western countries cannot just sit back and allow China to dominate media without prepared responses.

The China Proclamation serves as a basis/foundation for pursuing initiatives toward the strategy listed above. The Declaration of Independence was crucial for rallying support amongst colonists and gaining international recognition for their cause. The China Proclamation can do the same. Non-profits supporting China-related causes should be coordinated and mobilized.

All recommended initiatives below, in and of themselves, are not all critical for all countries. What is essential is that, combined, they generate the desired outcome. With all initiatives, it'll be essential that there is very close coordination with other countries. Again, a key point is that the proclamation serves as the justification for each initiative.

Not too long ago, I gave a presentation to the Rotary Club of Washington, DC. Several of the members were retired CIA officers, and I received a question: "Isn't what you are trying to do regime change?"

Traditional regime change often spirals into lengthy state-building projects or involves military action or covert operations. Those fail and are not worth the loss of blood and treasure. Examples are Iraq, Afghanistan, and Libya. What I am suggesting is not armed regime change; we will not need to invest in nation-building. This is not about toppling a government, but about catalyzing a shift in behavior through strategic pressure and global unity.

A key hope is that China's National People's Congress will ultimately start to appreciate the concerns of citizens (or loss of money) and positively change or remove Xi Jinping with a new election or appoint another individual. Many China scholars will point out that Xi changed China's constitution so he can't be made to leave. This is not true. If you actually read the constitution, you'll note that under Article 63 the National People's Congress has the authority to remove him. I am not suggesting the imposition of democracy at gunpoint; institutions will not have to be rebuilt like we did in Japan and Europe. Literature and studies of regime change in those instances showed it was a mistake.

What I propose is not a revolution, but an evolution—a carefully orchestrated series of moves on the global chessboard that could reshape the international order without a single shot being fired.

* * *

Watching China respond to global actions for about thirty-eight years now, I expect they will issue a press release along the lines of: "China strongly condemns the use of misinformation by the United States and other countries in an attempt to diminish China's position as a global partner. This will permanently harm our relationship."

With a unified Western world participating, China will be neutered from punishing any one group. Joe Biden's administration was exceptional at crafting talking points and then having supporting news outlets puppet them. This should be done when the document is released to pre-empt and marginalize China's anticipated press releases. The difference is that this should be clearly communicated and understood.

There is no need to respond to China. No country should speak with China leaders. This should be a blackout period. Unlike when a series of Biden officials paraded into Beijing to meet with Chinese officials, America should now do exactly the opposite. This is the beginning.

Many China scholars and diplomats will say you cannot isolate and humiliate China. They will anticipate Russia and China becoming closer, as well as DPRK and a few countries in South Africa. This, they'll say, will result in another cold war. Given Russia's geopolitical and economic situation, I do not see this happening. If China wants to start a cold war, in the face of Western countries' unity, that'd be a historic mistake. China will double down and try to form alliances with friendly countries, for sure. Efforts should be made to anticipate this and address them before they arise. America must become proactive.

As for being able to humiliate China, we have only just started. Let's remember—China scholars and diplomats have a nearly perfect record of misjudging China.

INITIATIVE 2:
SANCTION SENIOR CCP LEADERSHIP

If Bingbing were a top CCP official (which she actually could be) and America sanctioned her and took her assets in America—then her entire Tsinghua University class would know about it and be outraged. They would ask why and how could America do such a thing? Then, they would finally read and connect the dots to the The China Declaration. Her reaction and that of her Chinese friends would be the same: "How could China and its leaders be so stupid?"

*　　*　　*

After Trump left office, China sanctioned twenty-eight members of his staff. Biden sanctioned thousands of Russians regarding the invasion of Ukraine. Ironically, both failed to have any impact whatsoever. The main reason for this is that those individuals did not have a meaningful amount of assets within the area where they were sanctioned. This is not the case with regard to the corresponding Chinese. The CCP's elite, unlike their American counterparts, have woven their wealth into the fabric of Western economies, creating a vulnerability ripe for utilization.

There are ninety-nine million communist party members in China. The CCP has always approached China's brightest, asking them to join. They often do and get subsidized tuition, job opportunities, and other small benefits early on. These are not all bad people.

The structure (and power) of China's political bodies, from top down includes:

- Standing Committee: 7
- Politburo: 24
- National People's Congress: 2,977
- CCP Leadership in Companies & Other Orgs: 4,000

With England, France, and other Western countries, America should sanction all these individuals and their families. National Congress Members have directly supported the atrocities laid out in The China Proclamation (in Section VII). Additionally, according to China's Constitution, those serving in the National People's Congress have broad authorities and have not properly exercised them in the face of compelling evidence they should have done so. They must be held accountable.

Board members and CCP leaders within China-owned firms and 500 of its largest private firms are senior leaders in the party. They have influence within the CCP leadership. They are not bystanders.

Freezing their assets and those of their family members will be essential. Very often, assets such as homes are put in the name of family members. Children at universities should be given a few weeks to leave the country. Children of high-placed party members must not be permitted to benefit from Western education.

This move would send shockwaves through the upper echelons of Chinese society. Imagine the panic as the elites scramble to secure their wealth, their children's futures suddenly uncertain. It's a calculated gambit, designed to create fissures within the CCP's seemingly impenetrable facade.

Expectations are that some universities will protest the sending home of Chinese students. Over the past ten years, Chinese have donated over $1 billion to universities. China never gives money without realistic expectations of what they will receive in return. I am not typically in favor of investigations, but in this case, one may be warranted. Which universities received money and what did they promise China in return? I suspect that a number of universities will be shamed. Without a doubt, some universities align their priorities toward "common prosperity" goals as a means of attracting donations. This is a code phrase for supporting China. Consideration should be given to stop the federal funding of these universities.

Because there may be students who are noble, despite their parents, an exception application should be prepared prior to the execution of this initiative. Additionally, legislation should be aimed at prohibiting children of CCP members from attending American universities. This should be done separately, with consensus of all Western countries. The CCP is using America's educational system; when Chinese students arrive in America, they do not magically become capitalists or more respective of human rights.

Mostly, while at US education institutions, Chinese stick together, and speak Chinese. They continue to use their preferred social media apps and communicate with their friends and family via WeChat. Beliefs sometimes change. The most common outcome is that they remain loyal to their country and culture but maybe not to the country's regime.

We have to get the CCP's presence out of the educational system. This will again be another clear message to Chinese citizens that Western countries do not admire nor tolerate CCP members. There are about 900,000 Chinese students studying abroad. Sending the children of CCP members home will be a good concerted start to Chinese understanding that being a CCP member is *not* a good thing. This is an unknown concept in China that Western countries must teach CCP members and their children. Until now, they have only received benefits from being a CCP member or relative.

Those sanctioned members with the largest assets in Western countries should be advertised, at least in a press release. Ideally, there should be a leaderboard online that lists names by assets, both in English and Chinese. Chinese in China will want to keep track of this information. This should be encouraged and facilitated—even the Chinese government might find such wealth suspect.

America should use the money collected from sanctioned CCP members to fund non-CCP students to study in America. As previously mentioned in the Boxer Protocol, America received payments from China. After a while, America declined to receive all the payments with the condition moneys be channeled toward education. This led to the creation of China's leading university, Tsinghua University. A press release mentioning this to Chinese should be made. This should be spaced out so that another news cycle can be captured. Once again, America is helping to educate Chinese.

Welcoming non-CCP students reinforces a clear message: America loves the Chinese people, but not the CCP. Communication should go out to universities that these students should be warmly welcomed. This cannot be overly emphasized. A new generation of Chinese hostile to America is the last thing needed. In the world ahead, we are all stakeholders, together.

Western countries combined, doing such things, can start to have an impact to help Chinese appreciate the nature of their leaders. Chinese are hugely skeptical about Party bosses and how they earn their money. Seeing assets seized and students kicked out of Western universities will be well-received by Chinese.

Western countries will be sanctioning roughly 10,000 people or 0.01 percent of all CCP members. This is a targeted exercise as it is a comparatively small group of individuals. This initiative is not about punishing the Chinese people, but about surgically excising the cancer of corruption and oppression that has metastasized within the CCP's leadership.

There a number of press releases that can be put out justifying the action. "In response to China sanctioning Trump administration officials and because of the factual China Proclamation document, America is forced to express dissatisfaction with the dictatorial actions of President Xi. They have harmed Chinese citizens and countries globally."

*   *   *

It is expected that China will rebel against all Western countries with stern press releases. But, this is not a declaration of war and thus cannot justify a military response. The likely consequences are internal China aggravation. China is not in an economically strong position to the point it can do something harmful to all Western countries.

China will likely start to sell more US Treasuries. In 2013, they held $1.3 trillion and this has declined to about $760 billion. This should be monitored. If China starts to sell too much of their holdings, the Trump administration should jump to Initiative #6. Otherwise, China might feel the entire world is against them and decide it has nothing to lose by now moving on Taiwan. It sounds counterintuitive, but I'd station an extra fleet near the Taiwan Strait to discourage any such considerations.

INITIATIVE 3:
CLOSE EMBASSY & CONSULATES

Stop and ask yourself: "Would you keep letting someone into your house if they continually stole stuff?" No, you wouldn't. And yet, America continues to let China into America even though China keeps stealing IP. Everyone must start to have some common sense. China's Embassy and consulates must be closed in America. Before doing that, we should close our own in China.

After hearing about The China Proclamation and the sanctioning of senior CCP leaders, Chinese citizens may still not be paying attention. However, when Chinese see countries physically closing their embassies and consulates, they will start wondering—"What's going on?"

The older generation will relate to this on a deeper level, perhaps remembering stories about the eight nation alliance leaving after defeating China during the Boxer Rebellion. Offices closing will create fear within China leadership that countries are indeed leaving China and that they are abandoning it economically. The reality is different; no country has done anything economically to China, nor have companies been instructed to change their behavior whatsoever.

Bingbing and her ultra wealthy friends will see this and wonder with consternation— "What did the government do now? What's gonna happen next?"

*   *   *

China constantly tramples on companies' intellectual property rights and it has persistently maintained an unbalanced business environment, favoring domestic enterprises over foreign ones. Its capital controls are highly restrictive and outdated. But, it is not America's duty to facilitate US companies' operations in China, nor to shield them from legal entanglements. The responsibility lies squarely with the Chinese government, if it wishes to attract foreign investment. If US businesses require hand-holding in China, they would be wise to seek private assistance.

While those in the US Embassy might disagree, the facts are—its efforts have been ineffectual. They do not add value to American citizens or help protect America. Advocacy by the Embassy has never come close to justifying the expense of maintaining its physical presence in the country.

The US Embassy in Beijing, costing over $500 million, has become an anachronism in the age of the Internet and video conferencing. This architectural behemoth, once a symbol of diplomatic prowess, now stands as a monument to bureaucratic inertia.

Most people don't even know who the US Ambassador is. Visa processing for Chinese going to America rarely involves face-to-face interactions; they don't even allow Chinese citizens to speak with visa processing staff (the apex of bureaucratic arrogance I've personally experienced). The inconvenient interview in Guangzhou for Chinese getting a green card after marrying an American can be done via

Zoom. Passport renewals for US citizens can be handled via FedEx in Washington, DC, if necessary. Other services can be outsourced to US financial institutions in China. So, for example, you can have your passport delivered to a Citibank branch.

There are nearly 80,000 staff in the US State Department. Emptying the Embassy in China is a good start for a rationalization plan. Build a real visa app, outsource processing inquiries to the Philippines, if needed. Visa application fees should more than cover any processing expenses, including staff and rental.

I know there will be push back from State and Embassy officials and the CIA. So, let me elaborate on the reasoning why this is not a big deal:

The Embassy and CIA in Beijing have failed America. When COVID hit, we never heard about the Embassy or NSA/CIA calling Washington about a virus on an airplane heading toward America. The Embassy is supposed to help Americans in distress within China; I have factual evidence they have failed in this area too.

Historically, the CIA's effectiveness in China has been checkered at best. I remember around 2011 when China rounded up CIA informants and assets—they were executed or imprisoned. Estimates were about thirty such individuals were killed. In 2015, CIA officers and staff were removed from the Beijing Embassy as a result of a massive data breech at the Office of Personnel Management. The CIA/NSA's method of current day recruiting of teenage Mandarin language students in America is proof the agency is desperate and raises questions about its China-related competence.

After all of the above, does the CIA still use the Embassy and consulates to recruit spies in China? That'd be silly, since they are easy targets to track. There are only 22,000 foreigners living in Beijing, constituting 0.1 percent of the population. You can be sure that the China government allocates at least one person to track each of these foreigners. And, being posted at the Embassy is like putting a target on your back, making it easier for China to follow officers for counterintelligence efforts and technological surveillance.

I'm just saying that the national security interests of America will not be harmed by closing the Embassy, at least for a while. The Trump administration can readily deal directly with China officials—This should continue particularly during the execution of these initiatives. The Embassy should have zero communication with Xi's administration, unless instructed otherwise.

The big picture is that doing so will send a consequential message to China leadership and citizens. It says Western countries have lost respect for China. A historic document was signed, then countries left China. For those who worry—don't. We can always go back in. The point is to send a message (and make the State Department's operations more efficient during the process).

If the State Department is instructed to chat with China officials, they can use video conferencing or fly commercial to Beijing. Zoom exists. Airplanes exists. The

Internet exists. Preplanning must include things like better online visa processing and revising the State Department's antiquated website so people know where to go for various consular services.

Most importantly, we will need to help ensure Chinese staff within the Embassies and consulates can find jobs elsewhere. Near consulates, businesses and restaurants arose because of the people working there. This initiative should include payments in-kind to them, with the regrets of the US government. This will help create goodwill amongst the Chinese people. This consideration and approach to everyday Chinese is vital. Did the Embassy or CIA come up with this idea of being concerned about nearby businesses as part of this initiative? No, they would not because this is simply not the way they think.

*   *   *

The Chinese Embassy and four consulates in the United States should likewise be given a short period of time to close down. The US is shutting down China's presence in America. The China consulate in Los Angeles does not process visas online. A few weeks will give them plenty of time to create an online process. Otherwise, people can send visa materials directly to China for processing and pay a fee online.

The point of this is that we need to get China out of American cities so it is more difficult for Chinese officials to spy and harass Chinese in America, and as a clear response to China stealing from America. They are serving no productive function that adds value to China or its businesses, or—most importantly—to America. Any service Chinese need from China can certainly be provided by WeChat. After all, Chinese always say, "WeChat can do everything!"

Separately, a confidential telephone number to reach the FBI should be provided to Chinese who are being harassed by the CCP. Assurances of confidentiality will be of paramount importance. This should be mentioned in any related press release—that America wants to help Chinese when they are being harassed.

Powerful press statements relating to America's and China's consulates should be held with Chinese translations. Such press releases should refer to China's propensity to steal: "Guests who steal from those they are visiting are no longer welcome."

This isn't just about closing embassies; it's about rewriting the rules of international engagement. In one bold stroke, we're dismantling the outdated machinery of diplomacy and forcing a reckoning with an evil country.

* * *

China will take steps to speed up its nationalization to become self-sufficient—not having to buy or sell products to the rest of the world. They will likely stop selling rare earth elements and critical minerals, including ore and lithium-ion batteries. Preparations should be made early on to decrease reliance on China for these materials. They'll tariff key imported items such as agricultural products and seek other sources, leaving America farmers vulnerable.

It is also reasonable to expect China to take out its displeasure on Western companies. China leadership and CCP officials will do so out of irrational anger. Any China-based purchases from Western companies will be steered to other China-based suppliers, even if the cost is higher or products inferior. An embargo on iPhone sales in China could take place; this is the risk companies always knew existed when they decided to do business in China.

This initiative does not increase military action risk. But, moving a great many Western ships closer to China at this time would be icing on the cake. It would be an exclamation point to this initiative.

After a news conference and explaining the rationale behind this initiative, the administration should still refrain from speaking with China officials.

The consequences of this step are not severe. Unlike other steps, this one will be highly visible to Chinese and they'll start paying attention—finally.

I know that inevitably people will say, "Closing the US Embassy is crazy, just close part of it." These people are strategically misguided and operationally negligent. If their presence in the Embassy is always needed, they have been operationally negligent in not addressing such a weakness. The point is to send a clear cut message as part of a strategy. If it is truly necessary to have some people in the Embassy for national security reasons, then close the Embassy and simply leave guards behind. The critical staff can return after a week or so, unannounced to the press.

Also, if just the ambassador stays then he'll also need his secretary for access to his contacts. The ambassador will say he also needs his two key officers for data and news/source analysis. Oh, he'll need his chef and assistant for basic staff meals. Yeah, he'll also need his driver. And, security. And, just one housekeeper. Give a mouse a cookie . . .

At about this time, China will attempt to leverage its resources and friends in America to the maximum extent. Efforts will be made to place pressure on people to protest the actions of the United States. China will disburse funds to encourage influencers across the economy. Indirect efforts to influence those in the administration will occur. Third party charm offensives will occur. Through these efforts, China will endeavor to provide political wins to Trump in order to give him "face" to the American public. The public and administration should anticipate these endeavors, particularly the latter. There should be no confusion: The strategic objective of America has not yet been achieved.

INITIATIVE 4:
INCREASE CHINA IMMIGRATION

China's "Century of Humiliation" and its pride in its history have created a fierce independence and resistance from foreign influence. In a Tony Blair interview, Lee Kuan Yew strongly articulated Singapore needs the best intellectual capital in the world. Xi has taken the extreme opposite view with regard to China, believing it can and should be self-sufficient. This is seen in how China uses foreign talent, with a strong emphasis on temporarily learning from talent, not integrating them. This is pervasive across China. Foreigners are highly unlikely to ever obtain China citizenship after marrying a local. Not surprisingly, China does not publish statistics on how many passports it has issued to foreigners. The number is exceedingly low.

* * *

America and other Western nations should proactively attract the best and brightest from China. A coordinated branded program, used by all Western countries, should be launched so clear messaging is given. No Chinese who have ties to the CCP should be permitted under any circumstances.

In 2023, there were only 25,000 immigrants from China. This number should be significantly increased by President Trump, on humanitarian grounds until a new immigration law is approved. Essentially, individuals could apply online for a ten-year visa, provided they can demonstrate proof of exceptional ability and their potential value to America. This initiative would not only bolster America's talent pool but also send a clear message to China's leadership: the brightest minds in your country are choosing freedom over dictatorship.

The purpose of this initiative is to help America get even more of the brightest, and to attack China's nationalistic heart. This will spark questions as to the CCP's failure and legitimacy. After all, how can China grow economically if its smartest prefer to work and live elsewhere? I would suggest that funds obtained from sanctioning top China CCP leaders assets be used for this endeavor.

Increasing immigration to 125,000 annually, spread across the country, would not place undue pressure on America. This influx could provide a diverse solution to workforce shortages, particularly in manufacturing and even STEM-related fields, if they are highly qualified. Universities and companies can assist in their placement. We can give preference to their high tech/AI talent too.

Preference should be given to applicants who demonstrate proficiency in written English and have graduated from one of China's universities. (Those who write well in English can quickly learn to speak fluently.) Of course, CCP members or individuals with parents who are CCP members would be ineligible.

For President Xi, already grappling with a declining youth population, this would be a bitter pill to swallow. The irony of this initiative will be significant since China just recently stopped American adoptions of Chinese babies.

Nearly all American media outlets have repeatedly expressed concern about military-age Chinese citizens illegally crossing the border into America. These individuals are seeking better lives for themselves and their families. Typically, families send their most capable member first—someone who can work and establish roots. Is China inserting spies? Perhaps, but the greater concern should be China's theft of advanced technologies, like America's electromagnetic catapult system for aircraft carriers or F-35 designs now operational in China's military.

Media should be more concerned about all Chinese living in America who still have relatives in China. The CCP has a history of—formally or informally—blackmailing Chinese Americans by threatening their family members back home. This coercion can force individuals into compromising positions, making them vulnerable to CCP exploitation. This is standard operating procedure for the CCP and must be addressed as part of any broader immigration strategy. Consideration should be given to expedited immigration of families for those American Chinese who are being harassed.

A simple Zoom interview (by a capable interviewer) will go a long way toward ensuring applicants are suitable for immigration. Interviewers must be skilled at identifying honesty and truthfulness during these assessments. This process would not only safeguard national security but also ensure that America attracts individuals who genuinely align with its values and aspirations.

By opening doors to China's best and brightest while safeguarding against CCP influence, America can achieve two critical objectives: strengthen its own future while further seeding questions about Xi and the CCP's competence and legitimacy.

*   *   *

The biggest consequence of this Western country program will be resistance from Americans. Millions of illegal immigrants, the associated crime and cost, have tired Americans. Reassurance should be given to Americans, perhaps by introducing a few young bright Chinese.

Xi will be livid that America has an intentional policy to lure away China's best and brightest. However, this is no different than China's "Thousand Talents Program" that was launched by the Chinese government in 2008. The aim was to attract high-level experts from around the world by paying them large amounts of money. It was a form of bribery. In 2019, its name was changed. It was about this time that the US raised "concerns" about the program. Western countries are simply now doing what China previously did, but we are doing it better and more severely.

INITIATIVE 5:
SUE CHINA FOR FENTANYL DEATHS

This is a delicate and complex issue, but I believe there is no choice. Restrictive immunity is now largely being followed. A case can and should be made that the Chinese government has essentially enabled and allowed Fentanyl trafficking. With CIA/NSA assistance, clear evidence can be provided. Moreover, the trafficking of Fentanyl is a form of terrorism, as evidenced by about 150 deaths a day in the United States. This would fulfill the requirements of the US Foreign Sovereign Immunities Act (FSIA). However, this approach would likely be legally challenging, according to my discussions with legal scholars.

Many China experts believe the International Court should be used as an approach instead. Using the ICJ would face legal, jurisdictional, and practical barriers. Without going into detail, it is simply not an effective way to deal with this issue.

Instead, Congress should amend the FSIA to include new exceptions for cases related to public health (e.g., COVID) and drug trafficking. The groundwork for this initiative should begin earlier, if China begins selling of its Treasuries. We all know how long America's court system takes.

Will there be diplomatic repercussions? Expect some. But, let's be honest—China has been killing Americans. If you assume 200,000 Fentanyl deaths just last year, and each family gets $2 million, then China would have to pay $400 billion. But the total number of affected families is around 1,000,000. This would require a payout from China of an estimated $2 trillion USD. China currently holds about $775 billion in US. Treasuries. These should be held as collateral to ensure payment.

This move could send shockwaves through global financial markets. The mere threat of seizing Chinese-held treasuries could trigger a sell-off, potentially destabilizing the dollar. It's a high-stakes gambit that pits moral imperative against economic stability. I think it can and should be managed.

The point of this initiative is hopefully to succeed since it's the right thing to do. But a key goal is to diminish the legitimacy of China's senior CCP leadership. A barrage of these actions, if properly executed, is not just painful diplomatically. Citizens' awareness increases and business people become fearful. Americans now do not appreciate the magnitude of 150 people being killed each day. Chinese citizens haven't the slightest idea of what is going on. Like opium in China historically, Fentanyl is a political and social problem for America. Swift, decisive action and communication are mandatory.

*   *   *

Canada has the second largest impact from Fentanyl. European countries have experienced less severe crisis rates. So, broad Western support for this initiative may not be possible.

China will express diplomatic outrage and likely conduct a massive public relations campaign to portray itself as a victim of US legal overreach. Western countries need to get ahead of this before China responds. Any related press releases should be carefully crafted beforehand. China has not just been interfering with American society but actively killing its citizens.

China's leadership would first want to summon the US Ambassador to express "strong condemnation." America's response should be something along the lines of, "Sorry about that. Our Embassy has closed and everyone has already left." China might want to restrain American businesses or impose restrictions on diplomats in China.

China would also likely try to liquidate its financial positions, and perhaps sell other assets. This is the point when China will also get bolder militarily. For example, Taiwan is more at risk, as well as the Taiwan Strait.

If China has not done so by now, they will conduct non-military and/or non-traceable cyber strikes. These will include cyberattacks on financial systems, energy grids, and other networks. Besides coercive military posturing, great efforts will be placed on informational warfare. China will use all avenues to conduct social media campaigns and engage influencers.

The best response to such attacks is for America to publicize them. If America considers cyberattacks by China as an act of war, this should be clarified much earlier. My assumption is that America *will not and should not* respond using the advanced cyber tools in its arsenal at the expense of revealing capabilities. This is something worth contemplation prior to the execution of this strategy.

This initiative is not just about seeking justice for Fentanyl victims; it's also about fundamentally reshaping the rules of engagement with China. If China does something to America or its citizens, there will be an overwhelming retaliatory response. America will no longer sit idly by.

The amount of compensation China will pay is a negotiation worth participating in. This initiative may sounds crazy to many, almost like a tactical nuke. It must be orchestrated and bubble up over time so that China is not shocked when it hears about it. Simply put, if another country kills Americans, there will be repercussions. There will payback, literally or figuratively. Up until now there have been zero genuine repercussions. This is un-American, particularly since American intelligence services can tie this back to China.

INITIATIVE 6:
NO US ASSET OWNERSHIP

When doing my master's thesis, I learned you cannot trust China data. That was the case then and when I analyzed companies during my consulting days throughout China. Simply put—Chinese will get away with whatever they can.

America is permitting China's Public Company Oversight Board (PCAOB), the regulatory body, to oversee firms conducting audits of Chinese companies listed in America. In 2023, inspectors found deficiencies in 46 percent of the audits they reviewed. Chinese firms should face the exact same audit requirements American firms do when they list on US exchanges. If Chinese firms do not find this reasonable, then they can list elsewhere.

The Committee on Foreign Investment in the United States (SFIUS) has the authority to *recommend* a president block transactions. The US Department of Defense expanded a list of Chinese Military Companies which *might* lead to tightening scrutiny on any of their investments. This is not enough. As of 2023, Chinese entities owned about 384,000 acres of land in America.

There are several good examples of government-related companies owning land in security-sensitive areas. For example, a Chinese food producer, the Fufeng Group, purchased 370 acres of land near Grand Forks Air Force Base in North Dakota. Unless privately owned, all these should be sold. (As a footnote, China does not even allow the ownership of property in China by private US citizens.)

A more systematic, clear, and decisive approach should be legislated to counter China's current and likely future actions.

Congress should pass a law prohibiting the Chinese government and Chinese Communist Party (CCP) members from owning—directly or indirectly—any land, structures, or investment interests in US companies. Enforcement measures must be consistent and strict. This measure, a crucial safeguard against foreign influence, should be adopted by all Western nations. This legislation would serve as a financial firewall, protecting American assets from the insidious creep of CCP influence.

The scope of this law should include tangible assets, encompassing digital realms as well. It should include China-based social media platforms. Providing assistance to such companies should also be illegal and result in severe penalties. In other words, Apple would have to cease offering the TikTok app in its online store. This law would apply to WeChat, Xiaohongshu, Bililibi, and other platforms that are CCP owned. Any non-CCP owned company that does business in America must have its data in America and provide assurance the CCP cannot access it. This is a pain, but given China's track record, it is necessary. Clear guidelines will help Chinese companies avoid stiff penalties.

The US government should avoid instituting a complex web of laws that make buying and selling cumbersome. Instead, it should do exactly the reverse—ensuring a clear and efficient system so that viable private Chinese buyers can readily buy and sell.

As China grows geographically, so too must our approach to national security. This should become a broad law with regard to China, not simply adopted on a case-by-case basis by executive order. The stakes are too high for piecemeal solutions and it is challenging to predict what might otherwise happen. For example, when Americans thought TikTok would be disconnected, they immediately started opening accounts on WeChat and Xiaohongshu. The result is that young Americans are now using more China-based social media platforms. The risk of China having access to US citizen data and influence over American citizens is a basic threat everyone should understand. While everyone is focusing on TikTok, the same exact issue exists with WeChat, Xiaohongshu, and Bilibili. Instead of playing Whac-A-Mole, this should be addressed as part of this initiative.

Press releases must be intentionally crafted to send specific messages: "China has proven it cannot be trusted. It is not a good global citizen. America is treating China as China treats American companies in China." A key intention is to further diminish any remaining credibility the CCP might have amongst its business executives.

This isn't just about protecting American soil; it's about rewriting the rules of engagement. By closing the door on CCP-linked investments, while keeping it open for legitimate Chinese buyers, we're sending a clear message: America welcomes those who play by the rules of fair competition—but the days of the CCP's economic Trojan horse are over. It's a bold move that could reshape the global economic landscape, forcing China to confront the consequences of its actions on the world stage.

*   *   *

There will likely be no consequences from China having to divest any investment the CCP has in America. More than anything, they'll be surprised America didn't do this earlier. Creating legislation with an ability for Trump to make exceptions is a vastly better approach. Congress needs to get its act together to make things stricter, cleaner, and more consistent.

INITIATIVE 7:
REVOKE GRANTED
BIRTHRIGHT CITIZENSHIP

Earlier I mentioned counting newborn babies in US airports. There has been a constant stream of Chinese, and tens of thousands of CCP members, who have visited America to give birth. They did this explicitly so their child could eventually gain citizenship.

*   *   *

Without entertaining any discussions about the 14th Amendment, a press conference should be held to announce all births to Chinese having visited America over the past fifteen years will *retroactively* be *in*eligible for citizenship. The child will not be eligible to use their birth certificates to obtain obtain US citizenship. This is "because of the atrocities outlined in The China Proclamation perpetrated by the Chinese Communist Party leadership, all births by Chinese Nationals have no right to US citizenship."

This would be challenged if it got to the courts. The point is, Chinese will be deeply disappointed about all the time, effort, and money they put into arranging their child's birth. Hearing the above announcement, they will understand it is genuinely the fault of their government, not because of America.

The press in the United States will complain that retroactively revoking citizenship is not just unconstitutional, but also unjust. The following week, at another press conference, it can be announced that the policy has been changed. Instead, only children of CCP members who delivered their babies in the United States will be *in*eligible for citizenship.

Two news cycles will be captured by this subject and provide further reason for Chinese to tune into the situation. They will be pleased with the adjusted outcome. It is useful to remind Chinese citizens that China never offers citizenship to foreigners. China is a closed society, despite the dominance of Western clothing. America welcomes those Chinese who share America's values.

* * *

President Xi and his administration will have mixed reactions. This will cause embarrassment and some humiliation. If China is so great then why are so many people trying to secure citizenship elsewhere? He will feel anger at the United States for humiliating him.

Again, separating out the messaging from President Trump, except at appropriate times, will enable the voicing of a harsher, more inflammatory, tone. Another individual should head up these initiatives and conduct press briefings. Consideration should be given to such an individual making statements such as: "It is because China is headed by Xi, a dictator, that Chinese want to leave the country they so much love."

The truth is that a massive majority of Chinese will agree.

INITIATIVE 8:
NO TO TARIFFS

I do not advocate using tariffs as an initiative; however, I see the logic of President Trump using tariffs as a means of moving manufacturing back to America and changing country behavior. Otherwise, tariffs have never successfully been used to foster material behavior change in China—the type of change we are seeking.

Tariffs, while a blunt instrument in the economic toolbox, have proven to be more of a double-edged sword than a precision scalpel when dealing with China's entrenched policies. In particular, prior to pundits suggesting tariffs as a first option, people should understand China's resolve both in the past and currently when it comes to enduring such economic pain. Also, tariffs typically impact Chinese citizens, and that is strategically unwise. In short, tariffs do not take into consideration China and Chinese psyche as a result of past humiliation and national pride.

Similarly, sanctioning Chinese banks would harm commerce. Restricting access to the USD or locking China out of SWIFT would severely disrupt trade. Ultimately, this would hurt China's economy and the people. It would serve to alienate China and America rather dramatically hindering commerce which is a foundation to lives and relationships. I consider these as nuclear options that would not have the desired impact. Instead, I believe they would unite China against the world.

Also, imagine a scenario where China, backed into an economic corner, decides to accelerate its de-dollarization efforts, potentially triggering a seismic shift in global financial markets. I do not think this would occur, but, if it *did*, it could have ripple effects that could be catastrophic, not just for China and the US, but for the entire world economy.

This is in stark contrast to delegitimizing Chinese leadership as a means to an end. The end result of changing the behavior of CCP leadership would not be achieved through these broad and severe economic measures. Instead, we risk galvanizing support for the very regime we aim to pressure.

The path toward influencing China involves politics, economics, and psychology. There must be a good balance between pressure and persuasion, and leveraging the country's sense of humiliation and historic pride.

SECTION VI: CONCLUSION

Expect that China will issue all sorts of inflammatory statements indicating America is "interfering with China's internal affairs" and that this is "a serious violation of international norms." The difference is that there will be a coalition of partners, making it difficult for China to point its finger at any one country. To a limited degree, this will defuse any threats issued by China.

When it comes to China, worrisome issues are never-ending. Does it surprise you to hear that China buys about 90 percent of Iran's oil? Or that China has been actively hacking our telecommunications firms? What about China in Latin and Central America? What are you going to do about that? Let me emphasize again—This is the wrong question. People ask the wrong question all the time. Let's help everyone remember this.

The correct question is: *What should America's strategy be when it comes to China?* And: *What are the associated initiatives to successfully achieve the strategic objective?* People easily forget: Targeting or fixing "symptoms" does not produce success. You need a strategy. People often say, "Peace through strength" or "We need a whole of government approach" or "Leverage tools across the US government." The people who say these words are not only per-petuating the problem—they are part of the problem.

It is my great hope Western nations will act brilliantly and bravely. Those atrocities imposed by the CCP on its own people, and the world at large, cannot be allowed to continue. In recent history America consistently articu-lated "Strategic Competition" as its strategy. But, this has not produced any material results. It is time for more of an ideological competition.

What does the end-game look like? During the Boxer Rebellion (1899–1901), when all Western countries came together, China faced a humiliating military defeat. The goal now is similar insofar as China's leadership must be (to a good degree) humiliated and delegitimized, amongst themselves, and by China and global citizens. They should feel a potential inability to control society and their economy. Success is concrete evidence of China taking action to rectify at least some of the key issues outlined in The China Proclamation. I do not expect to see Xi apol-ogizing. But, initiatives should keep coming until clear capitulation occurs. Certainly there are more initiatives that can be executed and there are a host of other China issues that will need addressing as they arise.

*　　*　　*

Confucius once proclaimed, "The strength of a nation derives from the integrity of the home." Yet, in modern China, the sense of family as the nation's foundation has eroded, challenged by the rise of individualism, the decline of multigenerational households, and shrinking family sizes. This transformation of traditional family structures poses a significant threat to China's social fabric, undermining the very essence of what Confucius held dear.

The moral cultivation that Confucius championed begins at home, but its reach extends far beyond the family threshold. Insightfully, he posited a profound connection between personal morality and political governance, advocating for self-cultivation as a means to elevate society as a whole. This philosophy, once the bedrock of Chinese social order, now faces unprecedented challenges in a nation racing toward modernization.

China's trajectory has clearly diverged from the path laid out by Confucian philosophy. The nation that once revered filial piety and collective harmony now finds itself grappling with the consequences of rapid social change.

America and its families, however, present a stark contrast. Despite facing constant differences of opinion and the challenges of a diverse, pluralistic society, the nation remains remarkably resilient and unified. This strength, paradoxically, seems to stem from the very diversity and individualism that some fear may weaken China.

On January 20, 2025 (Taiwan Time), President Trump articulated this sentiment clearly to the cheers of those attending his victory rally the night before his inauguration: "We are one family, one glorious nation, under God." These were not empty words, but rather they echoed throughout the nation and in the hearts of citizens.

As China and America chart their distinct courses through the turbulent waters of the twenty-first century, the words of Confucius serve as a timeless reminder: the fate of nations is inextricably linked to the health of their most fundamental unit—the family.

America may have fewer people than China; America's past history may be shorter than China's. But America's sense of family is vastly stronger, as is its moral rectitude. In the grand theater of history, it is not the size of the cast nor the length of the play that determines its impact, but the strength of its central theme and the conviction of its players.

America is in a historically unique position, with the right cast and leading man, to address the challenges China presents.

SECTION VII:
THE CHINA PROCLAMATION

THE CHINA PROCLAMATION

The China Proclamation recognizes the universal truths that all people are created equal, and they are endowed with certain unalienable rights. *First* and foremost is Safety: One should face no fear of unprovoked harm from another human being. *Second* is Freedom: No one can stop another from doing as he wishes, so long as he does not infringe upon the rights of another. *Third*, Free Speech: No one can prevent another from freely sharing ideas and information. The *Fourth* is the Right to Vote.

To secure these rights, governments are instituted among men and women, deriving their just powers from the consent of the governed. Whenever a government consistently and dramatically tramples on these liberties, the governed possess the right to alter or abolish it and lay a new foundation to protect and foster their fundamental rights.

Should the international community witness a repressed people rendered unable to rebel against their oppressors, and that regime also has proven itself a menace to world peace and freedom, other countries have the individual and collective right to foment the changes that such a nation's embattled citizens cannot secure for themselves.

As leaders of the Free World, we are poised to advance the cause of peace. Human events sometimes command free nations to stand up, protest, and stop another power's egregious actions that abuse their own citizens. Ultimately, liberated people must elect their leaders.

Free countries have the right and duty to act. Inaction, short of military intervention, would be immoral. Diplomatic, cultural, and commercial pressure, and covert action are acceptable tools to transform tyrannies into free societies.

The above statements apply to the Chinese Communist Party (CCP), which has subjugated the Chinese people since 1949. The CCP has perpetrated massive atrocities against its own population since the dark days of Mao Zedong. When dealing with other countries, the CCP routinely lies, violates bilateral agreements, and breaks international law.

This list of the CCP's abuses and usurpations is long, but not exhaustive:

INTERNAL AGGRESSIONS

- China represses its citizens' free speech through technology by closing WeChat accounts of those who do not parrot the CCP's official line. (Source)

- China arrests and jails those who dissent from the CCP's opinions and edicts. (Source)

- "The CCP is committing crimes against humanity against Falun Gong practitioners through a billion-dollar state-sanctioned program of forced organ harvesting." (Source)

- The CCP places its financial interests above those of the Chinese people. (Source)

- In June of 1989 the People's Liberation Army stormed Tiananmen Square opening fire, crushing, killing, and arresting tens of thousands of Chinese protestors who were demanding greater political freedom. (Source)

- Within Xinjiang Province's internment camps, human rights abuses include widespread rape and torture. Chinese authorities have banished some 1.8 million people, primarily Muslim Uyghurs, to these off-limit internment camps. (Source)

- The CCP demolished Hong Kong's protective shield outlined in the Sino-British Joint Declaration, which was registered with the United Nations. The CCP's obliteration of its solemn agreement with Great Britain, Hong Kong's previous protector, crushed the rights of citizens to speak freely, publish at will, vote for their leaders, and remain safe within their homes and property. (Source)

- Since 1951, The CCP systematically has conducted a cultural genocide against Tibet and its people aiming to destroy their national and cultural identity. (Source)

- The CCP has forced women to endure 400 million abortions to protect its one-child policy. (Source)

- The CCP restricts its peoples' access to foreign news and opinions about the CCP's controversial actions inside and outside of China. (Source)

- China closely monitors and severely limits religious liberty. (Source)

- China's surveillance state monitors all citizens through electronic devices and other means. (Source)

- The CCP denies due process and hinders the development of an objective judicial system. It applies the law selectively, to benefit its allies and destroy its opponents. (Source)

EXTERNAL AGGRESSIONS

- The CCP permits the smuggling of fentanyl ingredients and other poisonous chemicals via Mexican drug lords into the United States. Each year, fentanyl kills more than 81,000 people in the US and has become the number one cause of death of Americans aged 18 to 45. (Source)

- The CCP breaks trade agreements including its 2001 signature on the pact that welcomed it into the World Trade Organization. (Source)

- The CCP violates its 2020 Economic and Trade Agreement with the US. (Source)

- The CCP steals and profits from American and Western intellectual property. (Source)

- The CCP threatens to invade Taiwan, intimidates the nation with belligerent military actions, and sanctions those who support the Republic of China's independence. (Source)

- The CCP helps Russian dictator Vladimir Putin's war on Ukraine through a "no limits" partnership and funds the Kremlin's unprovoked colonization of this independent nation by buying billions of dollars of Russian oil. (Source)

- China failed to warn the world promptly after the initial outbreak of COVID-19 in Wuhan. The pandemic contributed to the otherwise avoidable deaths of over six million people, with associated economic hardship. (Source)

Repeated attempts to remedy the above internal and external aggressions have failed:

- Forty countries worldwide have called on China to respect the human rights of Xinjiang's Uyghur community.

- Calls between US President Joe Biden and China's President Xi Jinping failed to discourage China from aiding Russia's invasion of Ukraine.

- The Chinese subjugation of Hong Kong received global condemnation but triggered zero change from China.

- China said it would ignore international court rulings regarding its behavior in the South China Sea, particularly regarding the Philippines.

Thus, today, we declare ourselves dedicated to bringing freedom from the influence of the CCP leadership to people in China and abroad.

We hereby support The China Proclamation:

SECTION VIII: EPILOGUE

The last time I saw Manman was October 1, 2023. We had spent the night together at the Courtyard Austin near the airport. She drove me and my two suitcases to the airport.

Last August 2024, I called Manman on my birthday. She did not answer. I know what that means. It means she can't, and I understand.

...

About Author

Brad Good, bestseller author of *The Control Center*, has lived and worked in the People's Republic of China since 1988. He brings rare on-the-ground knowledge of contemporary Chinese political, social, and cultural issues, and associated international affairs. Good has conducted business across the country and resided in various cities throughout China and Asia.

He holds an MBA and Master's in East Asian Studies from the University of Chicago and a BA from UC Berkeley. Good speaks fluent Mandarin and is a fifth-degree black belt in Shaolin Kempo Karate. He currently resides in Taipei, Taiwan.

www.ingramcontent.com/pod-product-compliance
Lightning Source LLC
Chambersburg PA
CBHW051516120626
46551CB00012B/951